# Collins

# Cambridge Lower Secondary

# Science

**PROGRESS BOOK 7:**
**TEACHER PACK**

**Series editor:** David Martindill
**Authors:** Aidan Gill, Emma Poole and Heidi Foxford

William Collins' dream of knowledge for all began with the publication of his first book in 1819.

A self-educated mill worker, he not only enriched millions of lives, but also founded a flourishing publishing house. Today, staying true to this spirit, Collins books are packed with inspiration, innovation and practical expertise. They place you at the centre of a world of possibility and give you exactly what you need to explore it.

Collins. Freedom to teach.

Published by Collins

An imprint of HarperCollinsPublishers
The News Building, 1 London Bridge Street, London, SE1 9GF, UK

HarperCollinsPublishers
Macken House, 39/40 Mayor Street Upper, Dublin 1, D01 C9W8, Ireland

Browse the complete Collins catalogue at
collins.co.uk

© HarperCollins*Publishers* Limited 2024

10 9 8 7 6 5 4 3 2 1

ISBN 978-0-00-867935-4

All rights reserved. No part of this publication may be reproduced, stored in a retrieval system, or transmitted in any form by any means, electronic, mechanical, photocopying, recording or otherwise, without the prior written permission of the Publisher or a licence permitting restricted copying in the United Kingdom issued by the Copyright Licensing Agency Ltd, 5th Floor, Shackleton House, 4 Battle Bridge Lane, London SE1 2HX.

British Library Cataloguing-in-Publication Data

A catalogue record for this publication is available from the British Library.

The questions, accompanying marks and mark schemes included in this resource have been written by the author and are for guidance only. They do not replicate examination papers and the questions in this resource will not appear in your exams. In examinations the way marks are awarded may be different. Any references to assessment and/or assessment preparation are the author's interpretation of the syllabus requirements.

This text has not been through the endorsement process for the Cambridge Pathway. Any references or materials related to answers, grades, papers or examinations are based on the opinion of the author. The Cambridge International Education syllabus or curriculum framework associated assessment guidance material and specimen papers should always be referred to for definitive guidance.

Series Editor: David Martindill
Authors: Aidan Gill, Emma Poole and Heidi Foxford
Publisher: Elaine Higgleton
Product manager: Catherine Martin
Product developer: Roisin Leahy
External Project Manager: Just Content Ltd
Development editor: Rebecca Ramsden
Copyeditor: Nick Hamar
Proofreader: Tanya Solomons
Cover designer: Gordon MacGilp
Cover illustrator: Ann Paganuzzi
Typesetter: PDQ Digital Media Solutions Ltd
Production controller: Sarah Hovell
Printed and bound by Ashford Colour Press Ltd.

We are grateful to the following teachers for providing feedback on the resources as they were developed: Dr. Rahul Sharma at IRA Global School, Mumbai, Mr Frank Akrofi and Mr Samuel Yeboah, Dániel Szücs at International School of Budapest, Ms Shalini Reddy at Manthan International School and Ms Sejal Vasrkar at SVKM JV Parekh International.

This book contains FSC™ certified paper and other controlled sources to ensure responsible forest management.

For more information visit: www.harpercollins.co.uk/green

# Contents

| | |
|---|---|
| **Introduction** | v |
| **Organisms and cells** | 1 |
| Self-assessment and reflective learning page | 5 |
| **Microorganisms and classification** | 7 |
| Self-assessment and reflective learning page | 13 |
| **Structure and properties of materials 1** | 15 |
| Self-assessment and reflective learning page | 20 |
| **Structure and properties of materials 2** | 22 |
| Self-assessment and reflective learning page | 27 |
| **Chemical changes and reactions** | 29 |
| Self-assessment and reflective learning page | 34 |
| **Energy and forces** | 36 |
| Self-assessment and reflective learning page | 42 |
| **Electricity and sound** | 44 |
| Self-assessment and reflective learning page | 50 |
| **The Earth and its atmosphere** | 52 |
| Self-assessment and reflective learning page | 59 |
| **The Earth in space** | 61 |
| Self-assessment and reflective learning page | 67 |
| | |
| **End of Year Test 1** | 69 |
| **End of Year Test 2** | 84 |

| | |
|---|---|
| **Mark scheme Organisms and cells** | **96** |
| **Mark scheme Microorganisms and classification** | **98** |
| **Mark scheme Structure and properties of materials 1** | **100** |
| **Mark scheme Structure and properties of materials 2** | **102** |
| **Mark scheme Chemical changes and reactions** | **104** |
| **Mark scheme Energy and forces** | **106** |
| **Mark scheme Electricity and sound** | **108** |
| **Mark scheme The Earth and its atmosphere** | **110** |
| **Mark scheme The Earth in space** | **113** |
| **Mark scheme End of Year Test 1** | **115** |
| **Mark scheme End of Year Test 2** | **119** |
| **Periodic Table** | **123** |
| **Glossary** | **124** |

# Introduction

This *Stage 7 Progress Teacher Pack* (and the *Stage 7 Progress Student's Book*) can be used to support the *Collins Cambridge Stage 7 Lower Secondary Science course* or to supplement your own resources. The *Progress Teacher Pack* contains

- nine End of Unit Tests offering practice questions to assess understanding of the Lower Secondary Science course
- two summative End of Year Tests
- student Self-assessment sheets for each of the End of Unit Tests
- mark schemes for each of the End of Unit Tests and for the End of Year Tests.

## How to use the Progress resources

This downloadable, editable and photocopiable Teacher Pack contains a range of End of Unit Tests that are designed to be valuable and flexible formative and summative assessment resources. They can be used to identify strengths and weaknesses and to pinpoint how future teaching should be adjusted to ensure all students make good progress.

The nine End of Unit Tests can be used as class tests or can be set for students to complete at home. They can be set at the end of a unit of teaching or can be combined to create a longer end of term test if appropriate.

Some of the questions in each End of Unit Test are written to address the Cambridge Thinking and Working Scientifically Learning Objectives:

- Models and representations
- Scientific enquiry: purpose and planning
- Carrying out scientific enquiry
- Scientific enquiry: analysis, evaluation and conclusions.

Each End of Unit Test is designed to be marked out of 20. Teachers may wish to set a time limit of 20 minutes.

The End of Year Tests assess objectives taught across the whole year. The style of the End of Year Tests is otherwise the same as the End of Unit Tests, with a mixture of question styles and question difficulties, as well as the inclusion of some Thinking and Working Scientifically questions. Questions are also set in the context of practicals where appropriate, ensuring that students have experience of answering questions on investigative work. These tests could be used for summative purposes as end of year examinations or as practice for students ahead of their examinations. The End of Unit Tests are separated out into the different science subjects, but the End of Year Tests cover a combination of the different science subjects.

The *Progress Teacher Pack* includes clear mark schemes for each End of Unit Test and for the End of Year Tests. These mark schemes contain notes on what should be seen for full marks to be awarded. They also set out how part marks can be awarded in a question where the full correct answer is not reached.

The student Self-assessment sheets give students the opportunity to reflect on their understanding. Students are given a list of statements about their understanding of the course content and are able to rank their understanding of the statement between 'I don't know', 'I need more practice' and 'I understand.' This will provide both students and teachers with a relatively quick way to assess the student's overall confidence with the content. They are also invited to rank their understanding of the Thinking and Working Scientifically content. Students are then invited to produce a written reflection to answer these questions: 'What went well in this topic?', 'What could you do better next time?' and 'What parts of the course could your teacher go through in a revision lesson which would support your improvement?'. There is also a space for teachers to make a comment about the student's understanding of the content.

Teachers can use the results of the End of Unit Tests and the students' Self-assessment sheets to help them in future lesson planning. For example, if many students struggled with work linked to the internal structure of the Earth, a teacher may wish to bear this in mind when planning their teaching of a related topic, such as seismic waves – teachers could, for instance, include starter activities recapping the earlier work.

## Key features: End of Unit and End of Year Tests

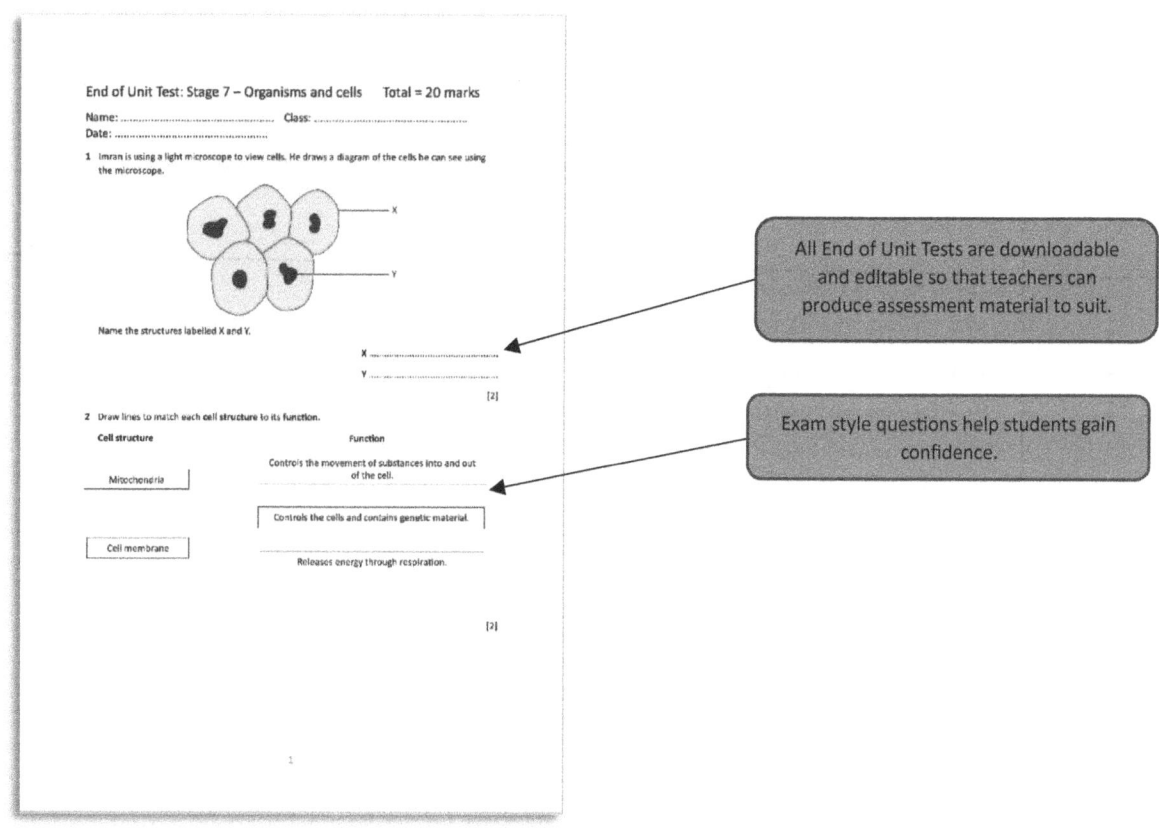

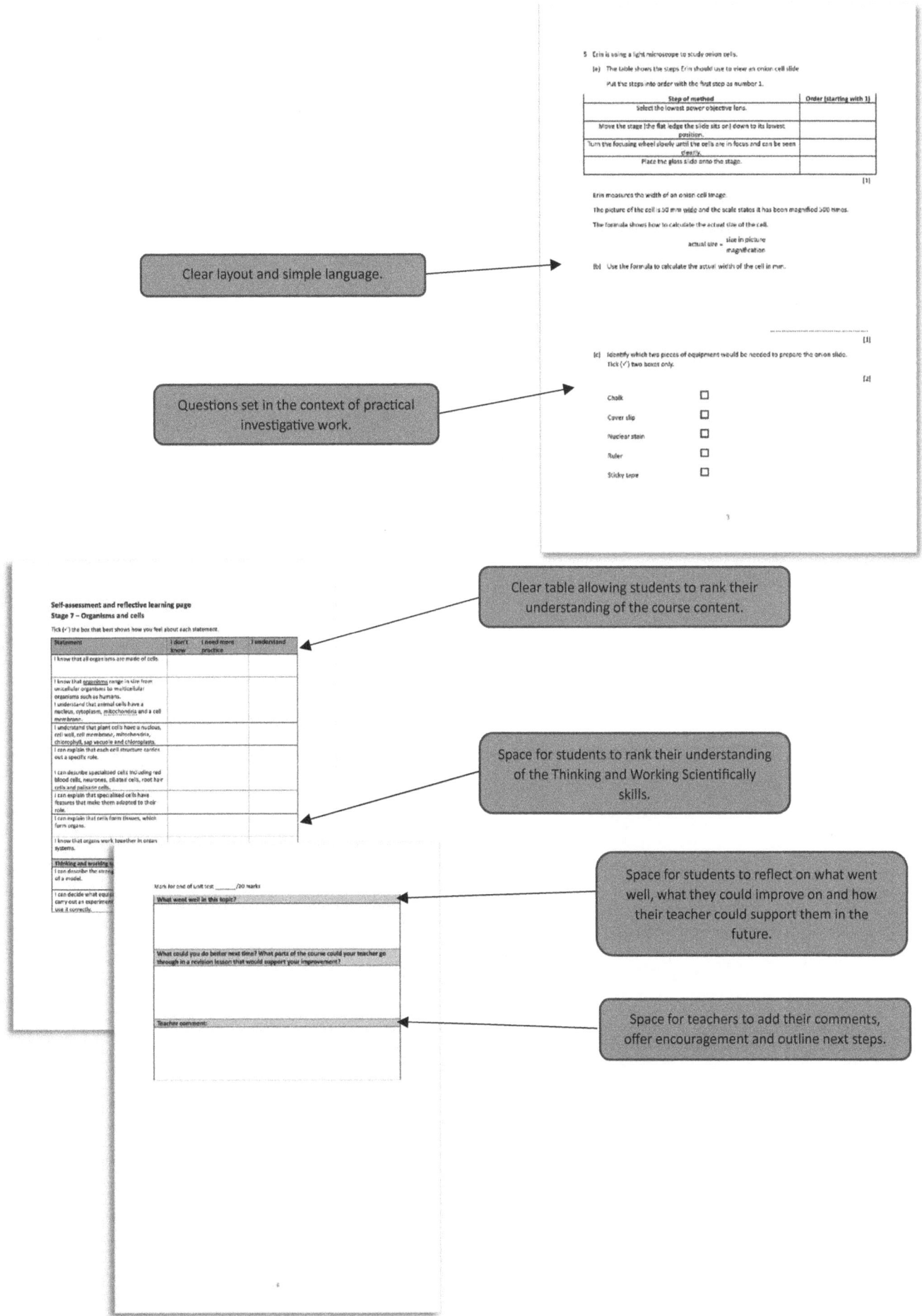

# Key features: Mark schemes

Clear indication of the correct answer and number of marks to be awarded.

Guidance given about how to award part marks

Information about which spec point the question relates to including Thinking and Working Scientifically statements.

Questions also linked to Assessment Objectives so that teachers can track which skills students need more support with.

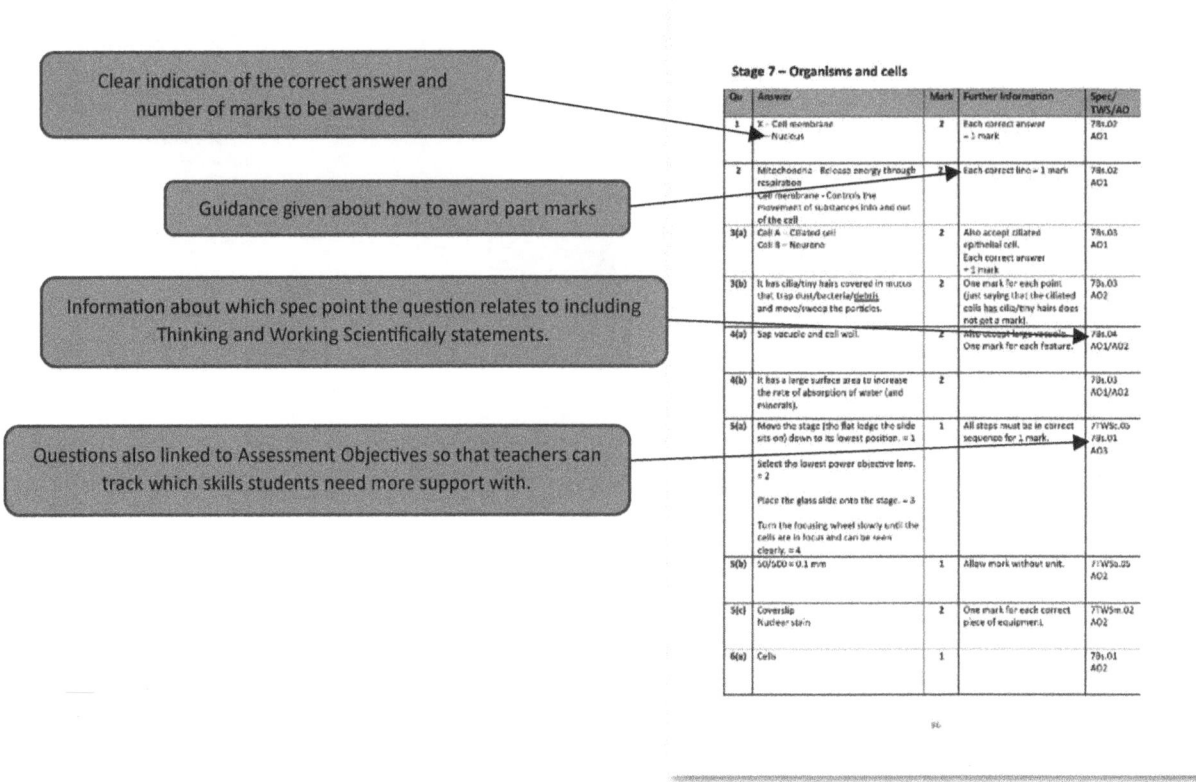

# End of Unit Test: Organisms and cells
# Total = 20 marks

Name: ................................................. Class: .................................................

Date: .................................................

1  Imran is using a light microscope to view cells. He draws a diagram of the cells he can see using the microscope.

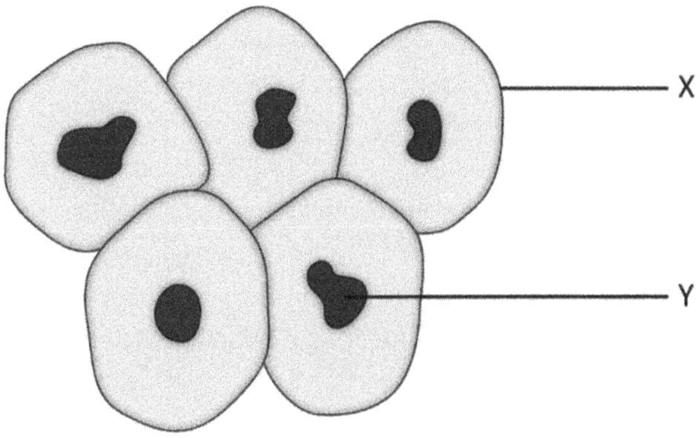

Name the structures labelled X and Y.

X ..............................................

Y ..............................................

[2]

2  Draw lines to match each **cell structure** to its **function.**

**Cell structure**

Mitochondria

Cell membrane

**Function**

Controls the movement of substances into and out of the cell.

Controls the cells and contains genetic material.

Releases energy through respiration.

[2]

**3** The diagram shows some specialised cells.

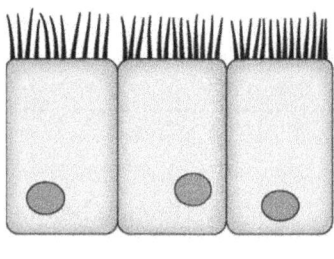
Cell A

Cell B

(a) Give the name of each type of cell.

**Cell A** ..................................................

**Cell B** ..................................................

[2]

(b) Explain how the structure of **Cell A** helps it to perform its function.

...................................................................................................................................................

...................................................................................................................................................

[2]

**4** Ziya is using a microscope to view a root hair cell from a cress plant. He draws a diagram of one of the root hair cells.

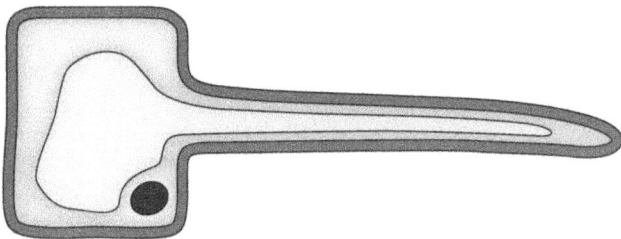

(a) Give **two** features of this cell that you would **not** find in an animal cell.

**1** ...................................................................................................................................................

**2** ...................................................................................................................................................

[2]

(b) Explain how the structure of the cress root hair cell is related to its function.

...................................................................................................................................................

...................................................................................................................................................

[2]

**5** Erin is using a light microscope to study onion cells.

(a) The table shows the steps Erin should use to view an onion cell slide.

Put the steps into order with the first step as number 1.

| Step of method | Order (starting with 1) |
|---|---|
| Select the lowest power objective lens. | |
| Move the stage (the flat ledge the slide sits on) down to its lowest position. | |
| Turn the focusing wheel slowly until the cells are in focus and can be seen clearly. | |
| Place the glass slide onto the stage. | |

[1]

Erin measures the width of an onion cell image.

The picture of the cell is 50 mm wide and the scale states it has been magnified 500 times.

The formula shows how to calculate the actual size of the cell.

$$\text{actual size} = \frac{\text{size in picture}}{\text{magnification}}$$

(b) Use the formula to calculate the actual width of the cell in mm.

.....................................................
[1]

(c) Identify which two pieces of equipment would be needed to prepare the onion slide.
Tick (✓) two boxes only.

[2]

Chalk ☐

Cover slip ☐

Nuclear stain ☐

Ruler ☐

Sticky tape ☐

**6** A flow chart can be used as a model to show different levels of organisation in living things.

> ............ > tissues > organs > organ systems >

(a) Complete the flow chart by naming the level of organisation before tissues.

[1]

(b) Identify one strength and one weakness of this model

[2]

**Strength:** ...........................................................................................................................

**Weakness:** .........................................................................................................................

(c) Write down which level of organisation the lungs belong to.

...................................................

[1]

# Self-assessment and reflective learning page
# Organisms and cells

Tick (✓) the box that best shows how you feel about each statement.

| Statement | I don't know | I need more practice | I understand |
|---|---|---|---|
| I know that all organisms are made of cells. | | | |
| I know that organisms range in size from unicellular organisms to multicellular organisms such as humans. | | | |
| I understand that animal cells have a nucleus, cytoplasm, mitochondria and a cell membrane. | | | |
| I understand that plant cells have a nucleus, cell wall, cell membrane, mitochondria, chlorophyll, sap vacuole and chloroplasts. | | | |
| I can explain that each cell structure carries out a specific role. | | | |
| I can describe specialised cells including red blood cells, neurones, ciliated cells, root hair cells and palisade cells. | | | |
| I can explain that specialised cells have features that make them adapted to their role. | | | |
| I can explain that cells form tissues, which form organs. | | | |
| I know that organs work together in organ systems. | | | |
| **Thinking and working scientifically** | | | |
| I can describe the strengths and limitations of a model. | | | |
| I can decide what equipment is needed to carry out an experiment and describe how to use it correctly. | | | |

Mark for end of unit test _____/20 marks

| **What went well in this topic?** |
|---|
| |

| **What could you do better next time? What parts of the course could your teacher go through in a revision lesson that would support your improvement?** |
|---|
| |

| **Teacher comment:** |
|---|
| |

# End of Unit Test: Microorganisms and classification
Total = 20 marks

Name: ................................................ Class: ..................................................

Date: ...................................................

1 Sureka is learning about living and non-living organisms.

(a) Circle **two living things** from the list.

**bird**       **fungus**       **rock**       **virus**       **water**

[2]

Sureka finds some living organisms in the school grounds. The diagrams show the organisms she has found.

stick insect       jumping spider

**Not to scale**

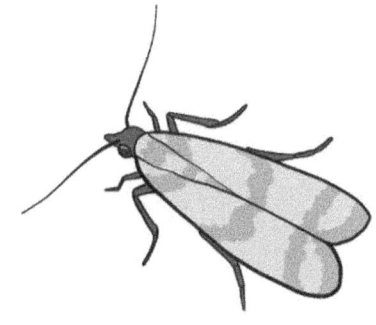

meal moth       black kite

Sureka makes a dichotomous key to help her class identify the organisms she has found.

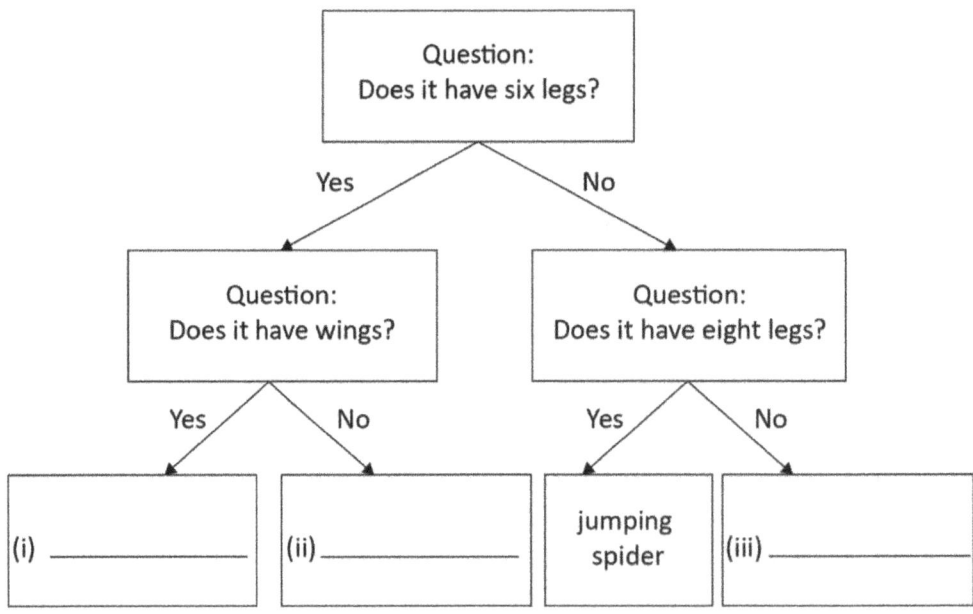

(b) Use the information in the dichotomous key to write the missing names of the organisms in the three empty boxes.

[1]

(c) The black kite is a medium-sized bird that belongs to the species *M. migrans.*

Which of these statements best describes a species? Tick (✓) the correct box.

A group of organisms that look exactly the same. ☐

An organism that can breed to produce infertile offspring. ☐

An organism that has been selectively bred to produce the best features. ☐

A group of organisms that can breed with one another to produce fertile offspring. ☐

[1]

2  The list shows six of the seven life processes that are found in all living things.

Movement

Reproduction

Sensitivity

Growth

Respiration

Excretion

....................................................

(a) Complete the list to show the missing life process.

[1]

(b) Write down what is meant by 'Respiration'.

................................................................................................................................

................................................................................................................................

[1]

(c) Living things can be classified as single celled or multicellular.

Draw a line from each **organism** to show if it is **single celled** or **multicellular**.

**Organism**

Yeast

Bacteria

Insect

**Singular celled or multicellular**

Single-celled organism

Multicellular organism

[2]

(d) Chicken pox is a disease caused by the varicella-zoster virus.

Scientists do not classify the varicella-zoster virus as a living organism.

Explain why the varicella-zoster virus is not considered to be a living organism.

................................................................................................................................

................................................................................................................................

[2]

**3** Ren is investigating the growth of bacteria in a colony that is growing on an agar plate. A colony is a group of bacteria that grow from a single bacterium.

The colony starts as one bacterium but doubles every 20 minutes.

(a) Calculate the number of bacteria in the colony after 2 hours.

...................................................
[2]

Ren uses the same bacteria to compare decomposition of bread kept in different conditions. He adds the same amount of bread and bacteria to each test tube.

(b) Ren sets up four test tubes as shown in the table below.

| Test tube | A | B | C | D |
|---|---|---|---|---|
| Condition | Warm and dry | Warm and moist | Cold and dry | Cold and moist |

Predict which test tube, **A**, **B**, **C** or **D**, would show the most decomposition of the bread after 7 days.

...................................................
[1]

(c) Explain why it is recommended to keep some foods in a fridge.

..........................................................................................................................

..........................................................................................................................
[1]

**4** Louis Pasteur believed that microorganisms could arise from non-living matter such as the air.

He put equal amounts of nutrient broth from boiled meat into two long-necked flasks.

Nutrient broth is the liquid made from boiling meat in water.

He left one flask with a straight neck. The other he bent to form an 'S' shape to trap any microbes in the air.

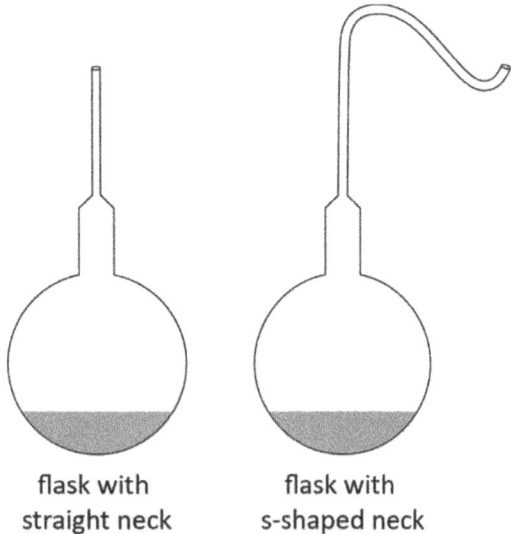

flask with straight neck    flask with s-shaped neck

Pasteur boiled the broth in each flask to kill any microorganisms in the liquid. The small tubes at the end of each flask were left open and the flasks were left in the same conditions for three weeks.

After three weeks Pasteur noticed that the broth in the straight-necked flask was discoloured and cloudy, but the broth in the S-shaped flask had not changed.

(a) Write down **one** variable Pasteur would have needed to control in his experiment.

.................................................

[1]

(b) Complete the sentences to write down what conclusions can be made from Pasteur's observations.

(i) The broth in the flask with the S-shaped neck did not spoil because

........................................................................................................................

........................................................................................................................

(ii) The broth in the flask with the straight neck became cloudy and discoloured because

........................................................................................................................

........................................................................................................................

[2]

Pasteur predicted that microorganisms in the air could get into the flask with the straight neck and cause the broth to spoil.

(c) Did Pasteur's results support his prediction. Explain your answer.

..................................................................................................................................

..................................................................................................................................
[1]

**5** In the 16th century people thought living things could come from non-living things. Francesco Redi didn't think this was true, so he used the scientific method to test the idea.

He set up three jars containing the same type and quantity of meat. One jar was open, one jar was sealed, and the other jar had a thick gauze covering the top to stop the flies touching the meat. The diagram shows the jars after two weeks.

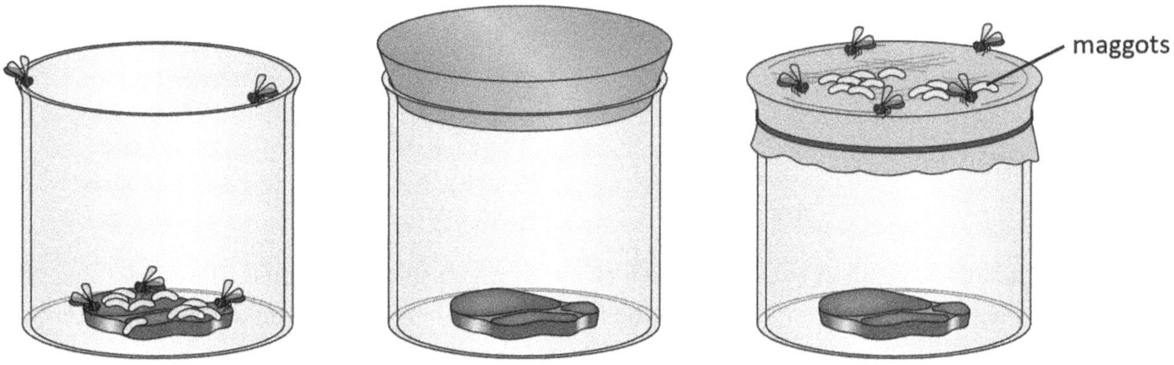

Redi's hypothesis was that flies laid eggs on the rotting meat, and maggots developed from those eggs.

(a) Do the results of Redi's experiment support his hypothesis?

....................................................
[1]

(b) Give a reason for your answer.

..................................................................................................................................
[1]

# Self-assessment and reflective learning page
## Microorganisms and classification

Tick (✓) the box that best shows how you feel about each statement.

| Statement | I don't know | I need more practice | I understand |
|---|---|---|---|
| I know that all organisms carry out seven life processes so that they can survive. | | | |
| I understand that organisms may be single celled or multicellular. | | | |
| I know that microorganisms are typically single celled. | | | |
| I can classify things as living or non-living. | | | |
| I know that a species is a group of organisms that can reproduce to produce fertile offspring. | | | |
| I know how to use a dichotomous key to classify organisms. | | | |
| I know that microorganisms play an important role in decay. | | | |
| **Thinking and working scientifically** | | | |
| I can classify organisms using keys. | | | |
| I can use formulae to represent scientific ideas. | | | |
| I can identify variables to control in an experiment. | | | |
| I can make a prediction using my knowledge of science. | | | |
| I can describe the accuracy of predictions, based on results. | | | |
| I can use the results of an investigation to decide whether or not the hypothesis has been supported. | | | |
| I can make a conclusion from observations. | | | |
| I know how Louis Pasteur used the scientific method to develop his ideas about microorganisms causing food to decay. | | | |

Mark for end of unit test _____ /20 marks

| **What went well in this topic?** |
|---|
| |

| **What could you do better next time? What parts of the course could your teacher go through in a revision lesson that would support your improvement?** |
|---|
| |

| **Teacher comment:** |
|---|
| |

# End of Unit Test: Structure and properties of materials 1
Total = 20 marks

Name: ................................... Class: ...................................
Date: ...................................

1 There are three states of matter.

   (a) In the box below draw how the particles are arranged in a **solid**.

      Use a circle to represent a particle. You should draw nine particles.

                                            [1]

   (b) Describe the way that particles move in a solid.

      ................................................................................................................................

      ................................................................................................................................
                                              [1]

   (c) Circle the state of matter in which the particles are moving very quickly in all directions.

              **solid**        **liquid**        **gas**

                                              [1]

   (d) Deep space is an example of a vacuum. Write down the meaning of the word vacuum.

      ................................................................................................................................

      ................................................................................................................................
                                              [1]

**2** A teacher tells her class that a chemical is corrosive.

Identify the hazard symbol for a corrosive chemical. Tick (✓) the correct box.

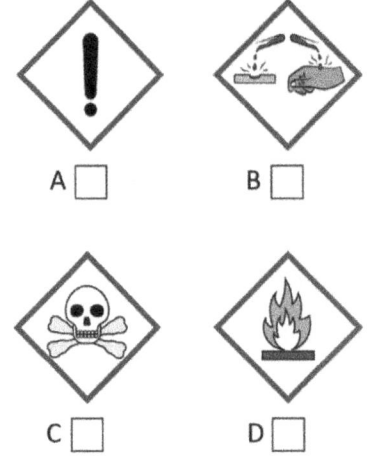

[1]

**3** Larry is given four solutions labelled **A, B, C** and **D**.

He tests solutions **A** and **B** with red and blue litmus paper and solutions **C** and **D** with universal indicator.

(a) (i) Complete the table to show the results of Larry's experiment using litmus paper.

| Solution | Colour with red litmus paper | Colour with blue litmus paper | Type of solution |
|---|---|---|---|
| A | red | .................................. | acid |
| B | blue | blue | .................................. |

[2]

(ii) Complete the table to show the results of Larry's experiment using universal indicator.

| Solution | Colour with universal indicator | Type of solution |
|---|---|---|
| C | green | .................................. |
| D | .................................. | weak acid |

[2]

(b) Larry knows he must work safely in the lab as he carries out experiments with acids and alkalis.

Explain **one** way that Larry should keep himself safe when working with acids.

...................................................................................................................................................

...................................................................................................................................................
[2]

(c) Larry says that universal indicator is a better indicator than litmus. Write down why universal indicator is a better indicator than litmus.

...................................................................................................................................................

...................................................................................................................................................
[1]

4 All substances have chemical properties and physical properties.

Draw lines to match each **property** to the correct **type of property**.

| Property | Type of property |
|---|---|
| Acidic or alkalinity | |
| | Chemical property |
| Melting point | |
| | Physical property |
| Good electrical conductor | |

[2]

5 Pheobe measures the melting points of three metals in her lab. The table shows her results.

| Metal | Melting point in °C |
|---|---|
| Iron | 1538 |
| Gold | 1064 |
| Copper | 1084 |

(a) Complete the graph showing Pheobe's results.

    (i)    Draw a bar to show the melting point of copper.

[1]

    (ii)    Write the missing label on the graph.

[1]

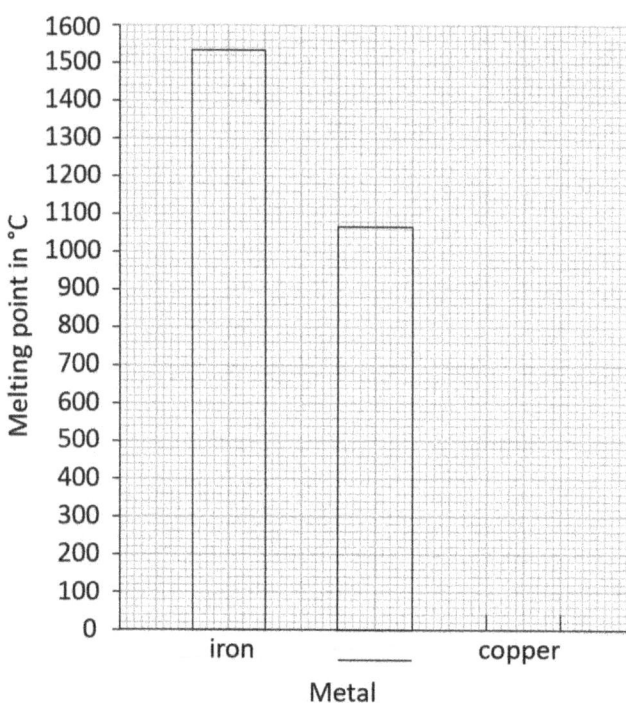

(b) Calculate the difference between the melting point of iron and the melting point of gold.

..................................................°C

[1]

(c) Suggest why it would be a good idea for another scientist to repeat Pheobe's experiment.

..................................................................................................................................

..................................................................................................................................

[1]

6 The particle model is used to show how particles are arranged and how the particles move in solids, liquids and gases. The particle model has strengths and limitations.

   (a) Write down **one strength** of the particle model.

   ............................................................................................................................................

   ............................................................................................................................................
   [1]

   (b) Write down **one limitation** of the particle model.

   ............................................................................................................................................

   ............................................................................................................................................
   [1]

# Self-assessment and reflective learning page
# Structure and properties of materials 1

Tick (✓) the box that best shows how you feel about each statement.

| Statement | I don't know | I need more practice | I understand |
|---|---|---|---|
| I can describe a vacuum as a space without matter. | | | |
| I can describe the three states of matter in terms of the arrangement, separation and motion of particles. | | | |
| I know that all substances have chemical properties. | | | |
| I know that all substances have physical properties. | | | |
| I understand that the acidity or alkalinity of a solution is a chemical property and is measured by pH. | | | |
| I know that indicators tell apart acidic, neutral and alkaline solutions. | | | |
| **Thinking and working scientifically** | | | |
| I can describe the strengths and limitations of a model. | | | |
| I can explain what each hazard symbol means. | | | |
| I can describe how to carry out practical work safely. | | | |
| I can present observations and measurements appropriately. | | | |
| I can interpret measurements and observations appropriately. | | | |
| I can evaluate experiments and investigations and suggest improvements. | | | |

Mark for end of unit test _____ /20 marks

| **What went well in this topic?** |
|---|
| |

| **What could you do better next time? What parts of the course could your teacher go through in a revision lesson that would support your improvement?** |
|---|
| |

| **Teacher comment:** |
|---|
| |

# End of Unit Test: Structure and properties of materials 2
Total = 20 marks

Name: ................................................ Class: ..................................................

Date: ..................................................

1   This question is about elements, mixtures and compounds.

   (a)  All matter is made up of **very small** particles. Name these very small particles.

   .................................................................................................................................
   [1]

   (b)  Steel is a useful **alloy** that can be used to make cars. What **type** of material is steel?

   Circle one answer.

                  element       mixture       compound      pure
   [1]

   (c)  The formula for carbon dioxide is $CO_2$. Explain why carbon dioxide is a **compound** and **not** an **element**.

   .................................................................................................................................

   .................................................................................................................................
   [1]

(d) The diagram shows the label from a bottle of mineral water.

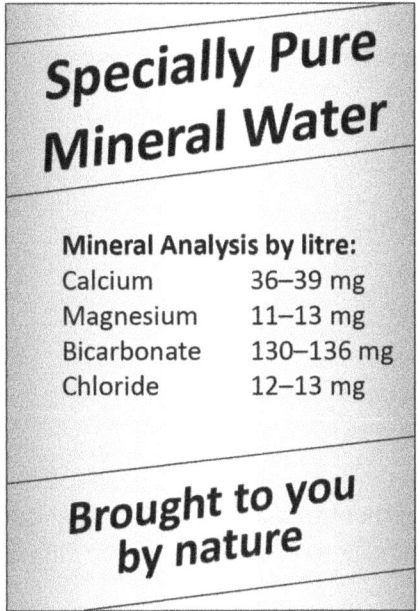

Write down why mineral water is **not** a compound.

..................................................................................................................................................

..................................................................................................................................................
[1]

**2** The Periodic Table at the end of the book shows names and symbols.

(a) (i) Use the Periodic Table to write down the **symbol** for the element sodium.

.....................................................
[1]

(ii) Use the Periodic Table to write down the **name** of the element that has the symbol Fe.

.....................................................
[1]

(b) Complete the sentences.

The Periodic Table shows all the chemical ................................................

Copper and magnesium are both found on the periodic table. Copper and magnesium both

belong to the ...................................... group.

[2]

**3** This question is about properties of metals.

(a) Copper is a brown coloured metal. It is used to make **electrical wires** and **saucepans** because it has useful physical properties. Tick (✓) **two** boxes to show the physical properties of metals like copper.

Good thermal insulator ☐

Good electrical conductor ☐

Conducts heat well ☐

Is brittle ☐

Has a very low melting point ☐

[2]

(b) Kevin carries out an experiment to find out which metal is the best conductor of heat. He has a kettle, a beaker and four spoons each made from a different metal.

Describe how he could carry out the experiment. Write down how he would know which metal was the best conductor of heat.

...........................................................................................................................................

...........................................................................................................................................

...........................................................................................................................................

[2]

**4** The diagram shows a **pure metal** and an **alloy**. Explain why alloys are harder than pure metals.

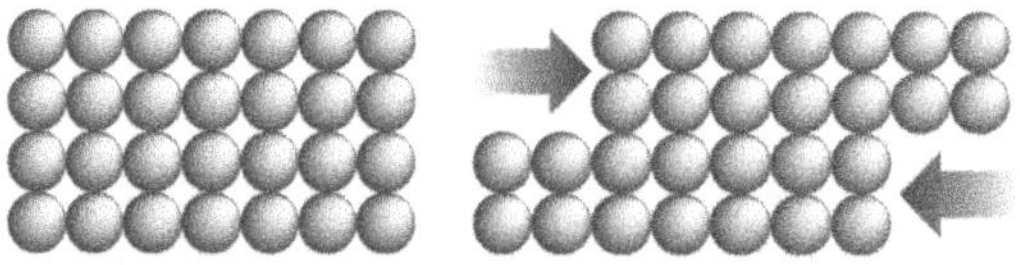

metal            alloy

...........................................................................................................................................

...........................................................................................................................................

...........................................................................................................................................

[2]

5  Look at the six diagrams below, labelled **A** to **F**. Complete the sentences to identify the letters that show elements, mixtures and compounds. Each letter can only be used once.

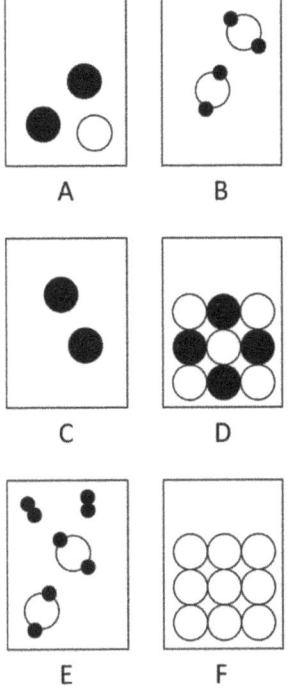

(a) The **elements** are ............................... and ....................................

[1]

(b) The **mixtures** are .................................. and ....................................

[1]

(c) The **compounds** are .................................. and ....................................

[1]

6  Copper can be combined with other metals to form alloys. The table below shows the percentage of copper in the alloy and its melting point.

| Percentage of copper in the alloy | Melting point in °C |
|---|---|
| 100 (pure copper) | 1083 |
| 85 | 1000 |
| 80 | 975 |
| 65 | 900 |
| 60 | 880 |

(a) Complete the graph to show how the melting point of copper alloys changes as the percentage of copper in the alloy changes.

   (i) Plot the value for the alloy that contains 60% copper.

[1]

   (ii) Draw a line of best fit.

[1]

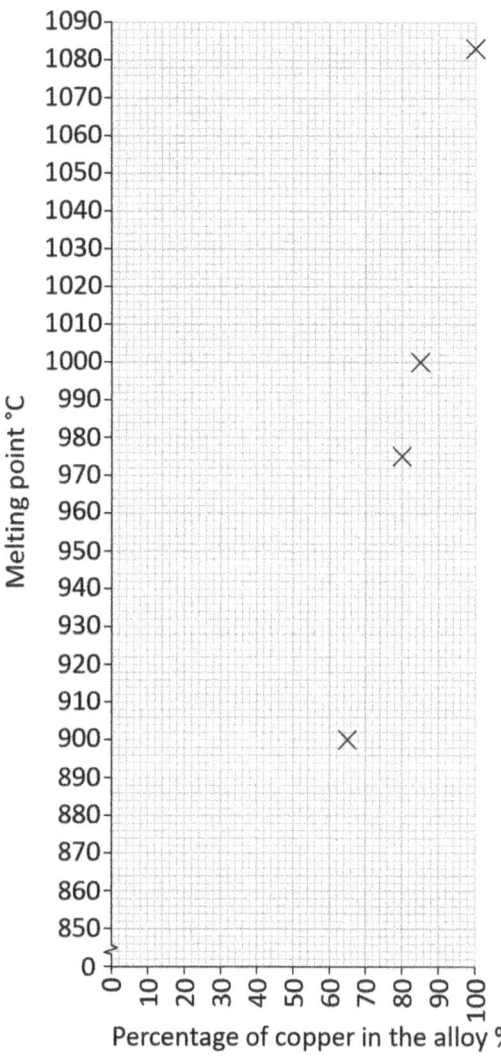

(b) Describe the trend shown in the graph.

..................................................................................................................................................

..................................................................................................................................................

[1]

# Self-assessment and reflective learning page
# Structure and properties of materials 2

Tick (✓) the box that best shows how you feel about each statement.

| Statement | I don't know | I need more practice | I understand |
|---|---|---|---|
| I understand that all matter is made of atoms, with each type of atom being a different element. | | | |
| I know the Periodic Table presents the known elements. | | | |
| I know that the two main groupings in the Periodic Table are metals and non-metals. | | | |
| I can describe the differences between elements, compounds and mixtures, including alloys as an example of a mixture. | | | |
| I can use the particle model to represent elements, compounds and mixtures. | | | |
| I can describe the physical properties of metals. | | | |
| I can use the particle model to explain the difference in hardness between pure metals and their alloys. | | | |
| **Thinking and working scientifically** | | | |
| I can use symbols and formulae to represent scientific ideas. | | | |
| I can plan how to carry out an investigation and consider the variables involved. | | | |
| I can record observations and/or measurements in an appropriate form. | | | |
| I can describe trends and patterns in results. | | | |
| I can present and interpret observations and measurements appropriately. | | | |

Mark for end of unit test _____/20 marks

| What went well in this topic? |
|---|
| |

| What could you do better next time? What parts of the course could your teacher go through in a revision lesson that would support your improvement? |
|---|
| |

| Teacher comment: |
|---|
| |

# End of Unit Test: Chemical changes and reactions
# Total = 20 marks

Name: .................................... Class: ....................................
Date: ....................................

1 This question is about chemical changes and physical changes.

Look at the list of observations.

**Changing colour**
**Temperature rise**
**Temperature remains constant**
**A liquid turns into a gas**
**Water freezes**
**Fizzing**

Complete the table to group the observations into **chemical changes** and **physical changes**.

| Chemical changes | Physical changes |
|---|---|
|  |  |
|  |  |
|  |  |
|  |  |

[2]

2 This question is about identifying different gases.

(a) Draw a line to link each **gas** to the correct **test for the gas**.

**Structure**          **Test to identify the gas**

Hydrogen               Relights a glowing splint

Oxygen                 Turns limewater cloudy

Carbon dioxide         Lit splint – squeaky pop

[2]

(b) Hydrogen is a gas and has this hazard symbol:

What type of hazard is shown in the diagram?

Tick (✓) the correct box.

Flammable ☐

Corrosive ☐

Hazardous to the environment ☐

Acutely toxic ☐

[1]

**3** This question is about different types of substance.

Look at these diagrams labelled A to C.

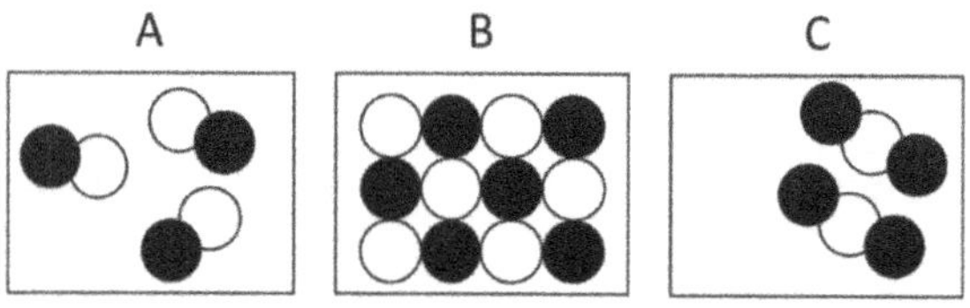

A B C

(a) What type of substance do all these models represent?

Tick (✓) the correct box.

Elements ☐

Compounds ☐

Mixtures ☐

[1]

(b) Draw how the particles would be arranged in a solid element.

[1]

**4** A student reacts colourless lead nitrate solution with colourless potassium iodide solution.

A chemical reaction takes place and a yellow solid called lead iodide appears.

(a) Write down the type of chemical reaction that has taken place.

.........................................................................................................................................

.........................................................................................................................................
[1]

(b) Write a word equation for the reaction between lead nitrate and potassium iodide.

.........................................................................................................................................

.........................................................................................................................................
[1]

(c) Explain why a yellow solid is formed in this reaction.

.........................................................................................................................................

.........................................................................................................................................
[1]

**5** A student heats iron and sulfur in a test tube to form one new substance.

(a) Give and explain one safety precaution the student should take when carrying out this experiment.

.........................................................................................................................................

.........................................................................................................................................
[1]

(b) Name the new substance formed in this reaction.

.........................................................................................................................................

.........................................................................................................................................
[1]

(c) The diagram shows how the particles are arranged before and after the reaction.

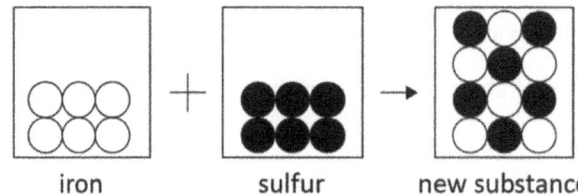

iron        sulfur        new substance

Explain how these diagrams show that a chemical reaction has taken place.

..................................................................................................................................................

..................................................................................................................................................
[1]

**6** Rachel has three test tubes labelled A, B and C. The test tubes each contain an equal volume of an acidic solution, but different concentrations of the same acid.

Rachel wants to know which test tube contains the most concentrated acid.

She adds Universal Indicator to each test tube. She then adds an alkali to the acid until the Universal Indicator turns green.

(a) (i) Write down the pH of the solution when it turns green.

..................................................................................................................................................
[1]

(ii) Write down the type of chemical reaction that takes place when exactly the right amount of acid and alkali react together.

..................................................................................................................................................
[1]

(b) To make the Universal Indicator turn green, Rachel has to add 30 drops of alkali to test tube A, 22 drops to test tube B and 15 drops to test tube C.

Design a table to record the results of her experiment. You should include the results of Rachel's experiment in your table.

[2]

(c) Compare the volume of alkali added to test tube A to the volume of alkali added to test tube C.

...................................................................................................................................................

...................................................................................................................................................
[2]

(d) Rachel used an equal volume of each acid. Give another way in which she can make sure that the experiment gives valid results.

...................................................................................................................................................

...................................................................................................................................................
[1]

# Self-assessment and reflective learning page
# Chemical changes and reactions

Tick (✓) the box that best shows how you feel about each statement.

| Statement | I don't know | I need more practice | I understand |
|---|---|---|---|
| I understand that matter is made of atoms with each different type of atom being a different element. | | | |
| I can use the particle model to represent elements, compounds and mixtures. | | | |
| I can describe how to use tests to identify hydrogen, carbon dioxide and oxygen gases. | | | |
| I can use observations to identify whether a chemical reaction is taking place. | | | |
| I can explain why a precipitate forms, in terms of a chemical reaction between soluble reactants forming at least one insoluble product. | | | |
| I can use the particle model to describe chemical changes. | | | |
| I can describe neutralisation reactions in terms of change of pH. | | | |
| **Thinking and working scientifically** | | | |
| I can use symbols and formulae to represent scientific ideas. | | | |
| I can plan a range of investigations and consider variables. | | | |
| I know the meaning of hazard symbols and consider them when planning practical work. | | | |
| I can describe how to carry out practical work safely. | | | |
| I can record observations and/or measurements in an appropriate form. | | | |
| I can present and interpret observations and measurements. | | | |

Mark for end of unit test _____ /20 marks

| **What went well in this topic?** |
|---|
| |

| **What could you do better next time? What parts of the course could your teacher go through in a revision lesson that would support your improvement?** |
|---|
| |

| **Teacher comment:** |
|---|
| |

# End of Unit Test: Energy and forces
# Total = 20 marks

Name: ................................................  Class: ...................................................

Date: ...................................................

1   The diagram shows a ball just after it has rolled off a table.

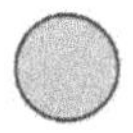

    (a)  Draw an arrow and label it to show the force acting on the ball.

    [1]

    (b)  Describe the effect of this force on the ball.

    ...............................................................................................................................
    [1]

2   This question is about energy transfers.

    (a)  Energy transfers take place in an electric kettle.

        Complete these sentences to state whether the energy transfer is useful or wasted.

        Energy transfer from electrical components to internal energy store of water is

        ...................................................

        Energy transfer from electrical components to outer case of kettle and air is

        ...................................................

    [2]

(b) Draw lines to match each **store of energy** to the correct **energy transfer**.

| **Store of energy** | **Energy transfer** |
|---|---|

Battery transfers energy …          … by light to surrounding.

Candle transfers energy …           … by sound to a person's ears.

Drum transfers energy …             … by electric current to torch bulb.

[2]

**3** The diagrams show two experiments, **A** and **B**, to measure the effects of gravity.

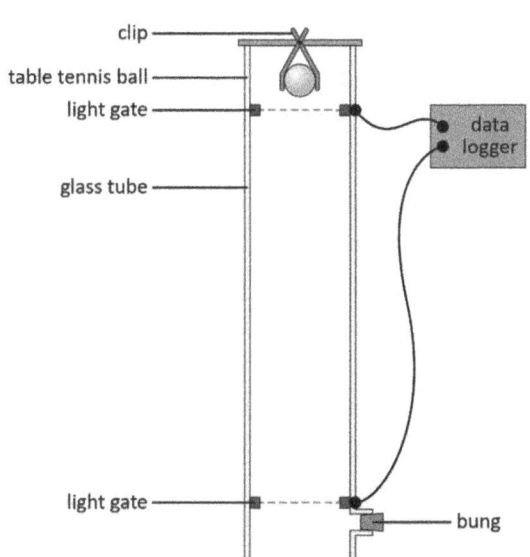
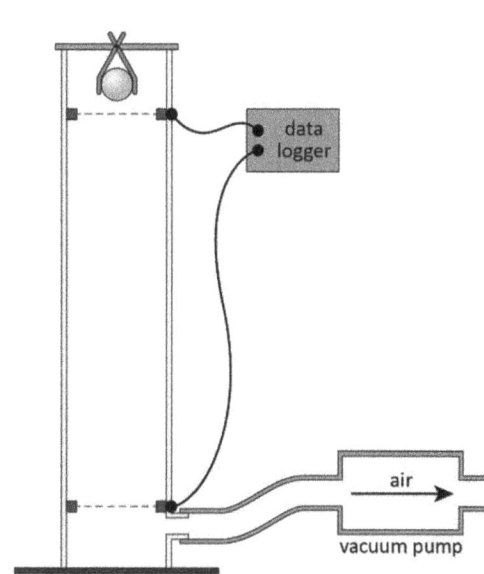

The two sets of apparatus are the same, except for a vacuum pump in **B**, which removes all the air from inside the tube.

To start the experiment, a table tennis ball is released by opening the clip at the top of each tube.

(a) In which experiment, **A** or **B**, is gravity the **only** effect that is observed? Explain your answer.

Experiment ...................................................................................................................

Reason ...........................................................................................................................

[2]

(b) Write down the effect **other** than gravity that can be observed in one of these experiments.

....................................................

37

**4** Chen investigates the energy transfers of a handheld vacuum cleaner.

Chen draws a diagram of the energy transfers.

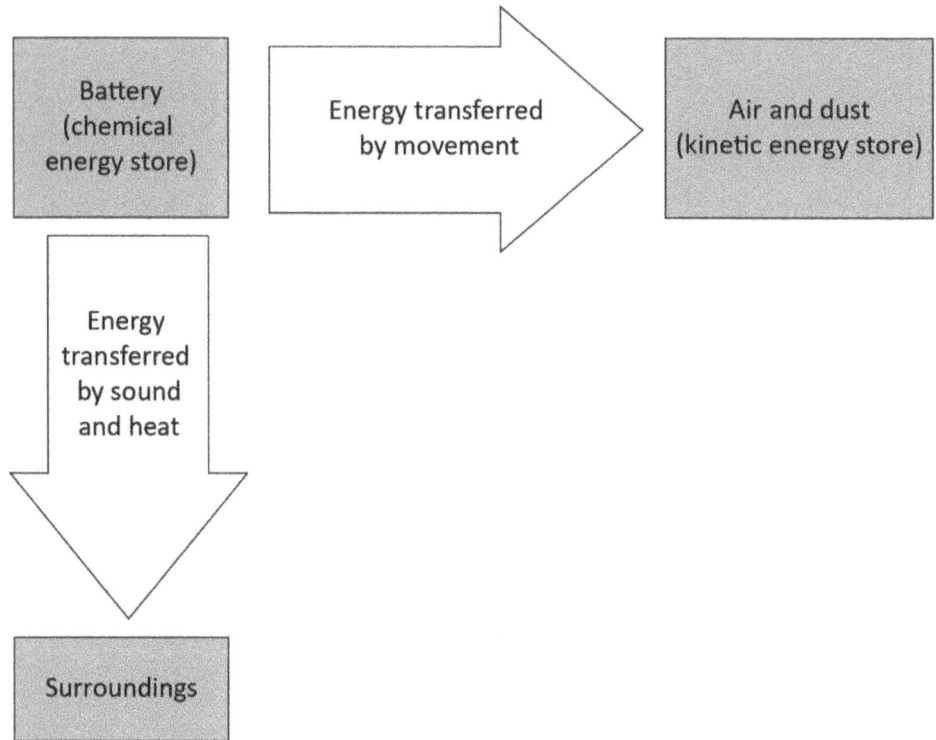

(a) Energy transferred to the surroundings is wasted. We say that this energy is:

Circle the correct answer.

              **dissipated**    **dissolved**    **missing**    **recycled**

[1]

(b) Describe the effect that this wasted energy has on the surroundings.

................................................................................................................................

[1]

5 The graph compares the efficiency of different electrical devices.

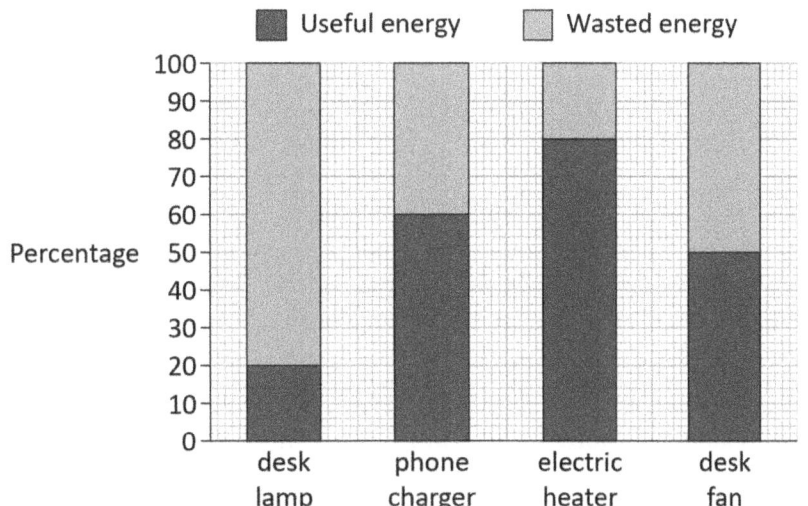

(a) Identify the device that wastes the largest percentage of energy.

.....................................................
[1]

(b) Write down the percentage of useful energy produced by the desk fan.

.....................................................%
[1]

6 The diagram shows some objects in the Solar System. (Sizes and distances are not to scale.)

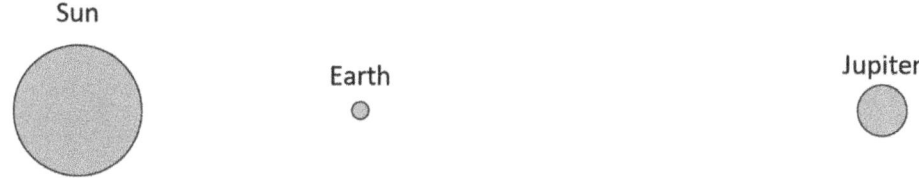

(a) Which of these objects do the other two objects orbit around?

Tick (✓) the correct box.

Jupiter ☐  Sun ☐  Earth ☐

[1]

(b) Explain why this object has this effect on the other objects.

....................................................................................................................

....................................................................................................................
[2]

**7** Rajiv investigates how stretched elastic stores energy.

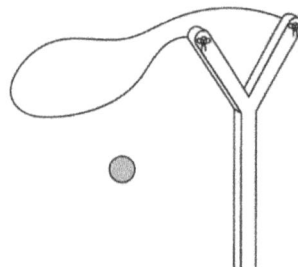

He places the ball in the catapult and stretches the elastic.

He releases the elastic and measures how far the ball travels.

(a) Predict the relationship between how much the elastic is stretched and how far the ball travels.

.......................................................................................................................................

.......................................................................................................................................
[1]

(b) Explain whether you think this experiment is safe or not.

.......................................................................................................................................

.......................................................................................................................................
[1]

**8** Fatima tests this hypothesis about gravity:

The time a dropped object takes to reach the ground does **not** depend on its mass.

She lists her apparatus.

- Safety goggles
- Tall stand with clamp
- Balls with different masses
- Data logger with light gates to measure time of the fall
- Measuring tape showing metres and millimetres

Fatima writes the following method.

1. Place ball at a height measured to be 2.0 m above the floor.

2. Release the ball so that it falls through a light gate.

3. Use the data logger to measure how long it takes the ball to reach the floor.

4. Carry out the same test with the other balls, testing each ball once.

(a) Identify the most important piece of equipment that has been missed out.

.................................................

[1]

(b) Write down **one** step to add to Fatima's method that could improve the results of her experiment.

..............................................................................................................................

..............................................................................................................................

..............................................................................................................................

..............................................................................................................................

[1]

# Self-assessment and reflective learning page
# Energy and forces

Tick (✓) the box that best shows how you feel about each statement.

| Statement | I don't know | I need more practice | I understand |
|---|---|---|---|
| I can explain that energy changes make things happen by transferring energy from a source to other objects or devices. | | | |
| I know that energy can be transferred by electricity, by sound, by light, by heating or by doing work (using a force to make an object move). | | | |
| I can explain that energy dissipates. This means that with every useful energy transfer there is always also some wasted energy, which is less useful. | | | |
| I understand that there is an attractive force of gravity between any two masses. | | | |
| I understand that the size of the force of gravity depends on the masses of the objects. | | | |
| I know that there are no particles and no air resistance in a vacuum. | | | |
| **Thinking and working scientifically** | | | |
| I can use symbols to represent scientific ideas. | | | |
| I can make predictions of likely outcomes for a scientific enquiry based on scientific knowledge and understanding. | | | |
| I can decide what equipment is required to carry out an experiment and describe how to use it appropriately. | | | |
| I can describe how to carry out practical work safely. | | | |
| I can describe trends and patterns in results. | | | |
| I can make conclusions by interpreting results. | | | |
| I can evaluate experiments and suggest improvements. | | | |

Mark for end of unit test _____ /20 marks

| **What went well in this topic?** |
|---|
| |

| **What could you do better next time? What parts of the course could your teacher go through in a revision lesson that would support your improvement?** |
|---|
| |

| **Teacher comment:** |
|---|
| |

# End of Unit Test: Electricity and sound
Total = 20 marks

Name: .................................................. Class: ....................................................

Date: ....................................................

1 Which diagram, **A**, **B**, **C** or **D**, shows the arrangement of particles in a sound wave?

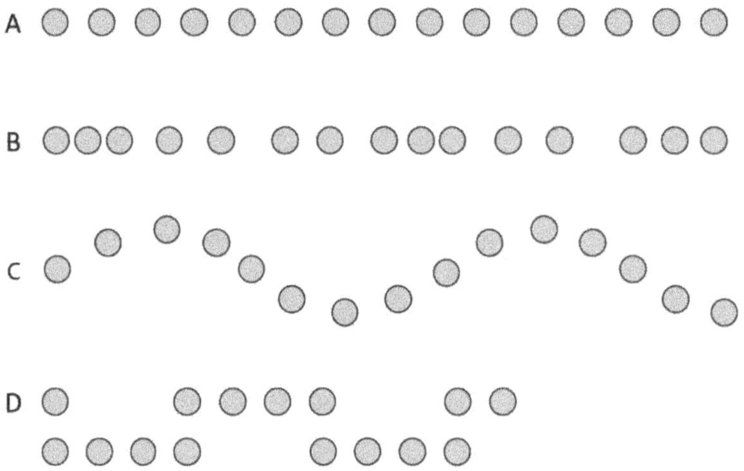

A ⊙⊙⊙⊙⊙⊙⊙⊙⊙⊙⊙⊙⊙⊙

B ⊙⊙⊙ ⊙ ⊙ ⊙ ⊙ ⊙⊙⊙ ⊙ ⊙  ⊙⊙⊙

C (wave-shaped arrangement)

D (two rows of circles with gaps)

Tick (✓) the correct box.

**A** ☐          **B** ☐          **C** ☐          **D** ☐

[1]

2 The diagrams show electrical components. Identify each component.

(a)

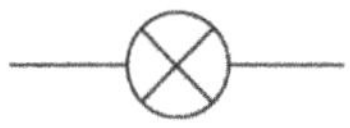

—⊗—

....................................................

[1]

(b)

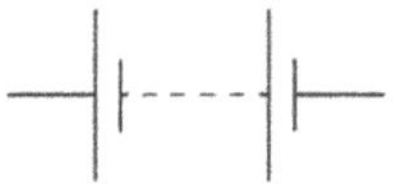

—|⊦- - -⊦|—

....................................................

[1]

**3** (a) Complete this sentence about materials and electric current.

An electrical conductor is a material that ...................................................................................

[1]

(b) Many electrical cables are made of copper wrapped in a thick layer of flexible plastic.

Complete the sentences about electrical cables.

Copper is an electrical ....................................................

Plastic is an electrical ......................................................, so it protects people from electric shocks.

[2]

**4** Safia walks next to a tall cliff. When her friend calls 'hello' to her once, she hears the word twice.

(a) Name the process that causes the second sound and explain why a second sound is heard.

.......................................................................

[1]

Later that day, Safia and her friend return to the cliff with a microphone and a recording device to measure the loudness of the sounds.

They produce a graph to show the loudness of the sound when one of them says a word once.

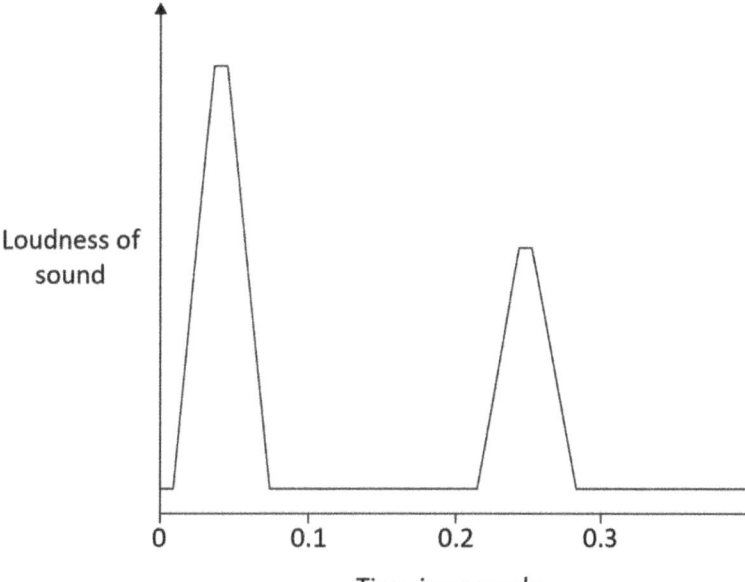

(b) What does each peak of the graph show?

................................................................................................................................................

................................................................................................................................................

[1]

**5** The diagram shows an electrical circuit.

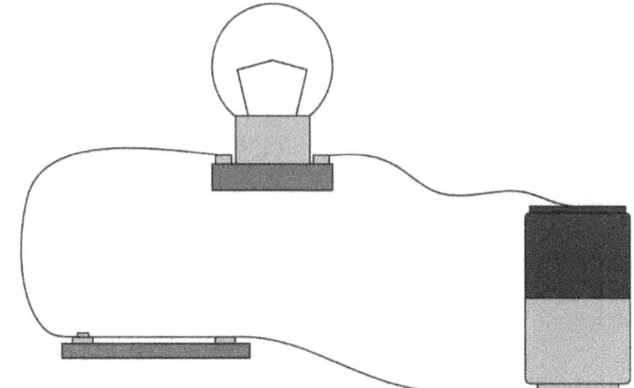

(a) Draw **three** arrows on the diagram (one on each wire) to show the flow of electrons.

[1]

(b) Write down what will happen to the flow of electrons if the switch is opened.

.................................................................................................................................................

.................................................................................................................................................

[1]

**6** James investigates the electrical circuit shown in the diagram.

The diagram also shows the current reading.

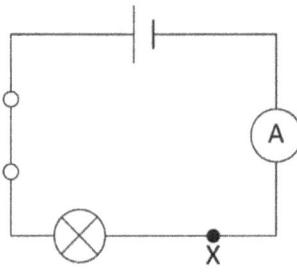 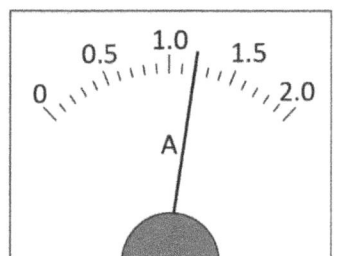

(a) Write down the current measured.

..................................................

[1]

(b) Predict what will happen to the current if a lamp is added to the circuit at point **X**.

.................................................................................................................................................

.................................................................................................................................................

[1]

**7** Blake investigates sound.

She uses the apparatus shown in the diagram.

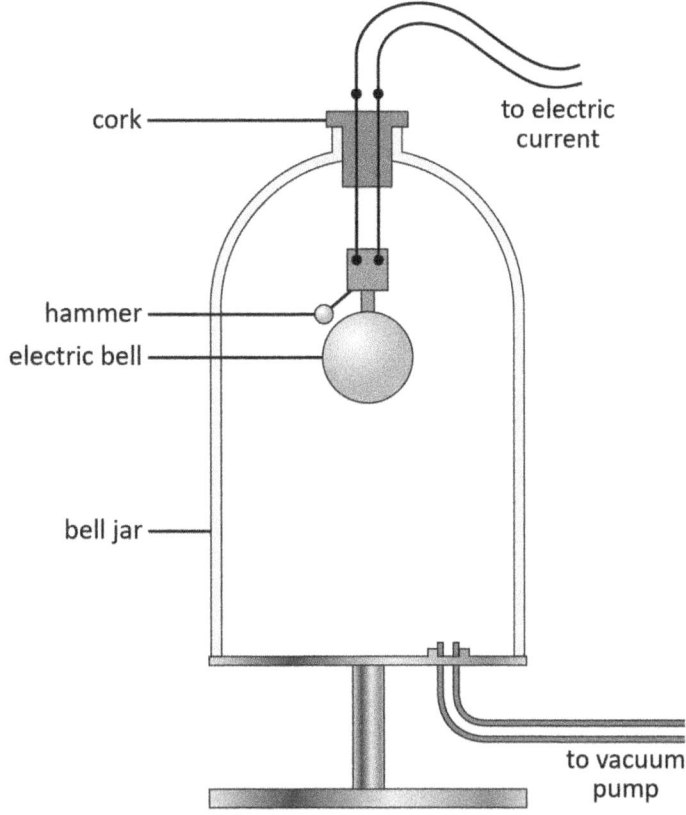

As air is pumped out of the container, Blake observes that the sound becomes gradually quieter until she cannot hear it.

Blake writes the following conclusion:

As the amount of air decreases, the bell vibrates more slowly until eventually it stops.

(a) Evaluate Blake's conclusion.

...................................................................................................................................

...................................................................................................................................
[1]

(b) Write your own conclusion.

...................................................................................................................................

...................................................................................................................................
[1]

**8** Pierre investigates the electrical circuit shown in the diagram.

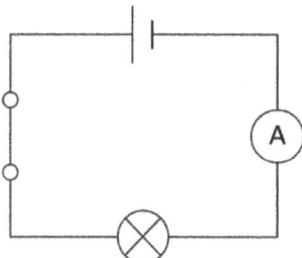

(a) Name the component used to measure the electric current.

.....................................................
[1]

(b) Pierre adds more cells to the circuit in series.

These are Pierre's results.

| Number of cells | Current in A |
|---|---|
| 1 | 0.3 |
| 2 | 0.6 |
| 3 | 0.8 |
| 4 | 1.0 |

Describe the pattern shown in the results.

..........................................................................................................................

..........................................................................................................................
[1]

(c) When Pierre adds a fifth cell, the lamp goes out and the current reading is 0 A.

Suggest an explanation for this.

..........................................................................................................................

..........................................................................................................................
[1]

**9** A microphone transfers the energy from sound waves to electric current.

Lily and Angelique investigate how the electric current produced in a circuit by a microphone is affected by the loudness of sound. They use a sound meter to measure the loudness in decibels (dB). The louder the sound, the greater the number of decibels.

They take three separate readings of current for each loudness of sound.

This table shows their results.

| Loudness of sound in dB | Current produced in A | | | Mean current in A |
|---|---|---|---|---|
| | Reading 1 | Reading 2 | Reading 3 | |
| 40 | 0.60 | 0.64 | 0.62 | 0.62 |
| 50 | 0.75 | 0.91 | 0.77 | |
| 60 | 0.92 | 0.92 | 0.95 | .................... |

(a) One of the readings is anomalous. Identify this reading.

Loudness ..................................... dB    Reading (1, 2 or 3) ..............................................

[1]

(b) Which mean current causes a sound with a loudness of 50 dB? Circle the correct answer.

...........................................................................................................................................

...........................................................................................................................................

          0.51 A         0.76 A         0.81 A         0.93 A

[1]

(c) Calculate the mean current for a loudness of 60 dB. Write your answer in the table.

[1]

# Self-assessment and reflective learning page
# Electricity and sound

Tick (✓) the box that best shows how you feel about each statement.

| Statement | I don't know | I need more practice | I understand |
|---|---|---|---|
| I know that sound is caused by the vibration of particles. | | | |
| I know that sound travels as a wave. | | | |
| I know that sound needs a medium to travel through so sound cannot travel through a vacuum. | | | |
| I can explain how echoes are made by sound waves reflecting. | | | |
| I can describe current as a flow of charge in a circuit. | | | |
| I know that cells provide energy to make current flow. | | | |
| I can explain that metals are good conductors because they allow current to flow. | | | |
| I can explain that most non-metals are good insulators because they do not let current flow. | | | |
| I understand that current is measured in amps using an ammeter. | | | |
| I know that the current is the same all round a series circuit. | | | |
| I can explain that adding cells to a circuit increases the current. | | | |
| I can explain that adding extra lamps to a circuit decreases the current. | | | |
| **Thinking and working scientifically** | | | |
| I can use electrical circuit symbols to represent components. | | | |
| I can make predictions of likely outcomes for a scientific enquiry based on scientific knowledge and understanding. | | | |
| I can decide what equipment is required to carry out an experiment and describe how to use it appropriately. | | | |
| I can describe how to take appropriately accurate and precise measurements. | | | |
| I can describe trends and patterns in results, including identifying any anomalous results. | | | |
| I can record sufficient observations and/or measurements in an appropriate form. | | | |

| I can make conclusions by interpreting results and explain the limitations of the conclusions. | | | |
|---|---|---|---|
| I can present and interpret observations and measurements appropriately. | | | |

Mark for end of unit test _____ /20 marks

| **What went well in this topic?** |
|---|
| |

| **What could you do better next time? What parts of the course could your teacher go through in a revision lesson that would support your improvement?** |
|---|
| |

| **Teacher comment:** |
|---|
| |

# End of Unit Test: The Earth and its atmosphere
# Total = 20 marks

Name: ................................................. Class: ...................................................

Date: .................................................

1 Look at the diagram.

   It shows the internal structure of the Earth.

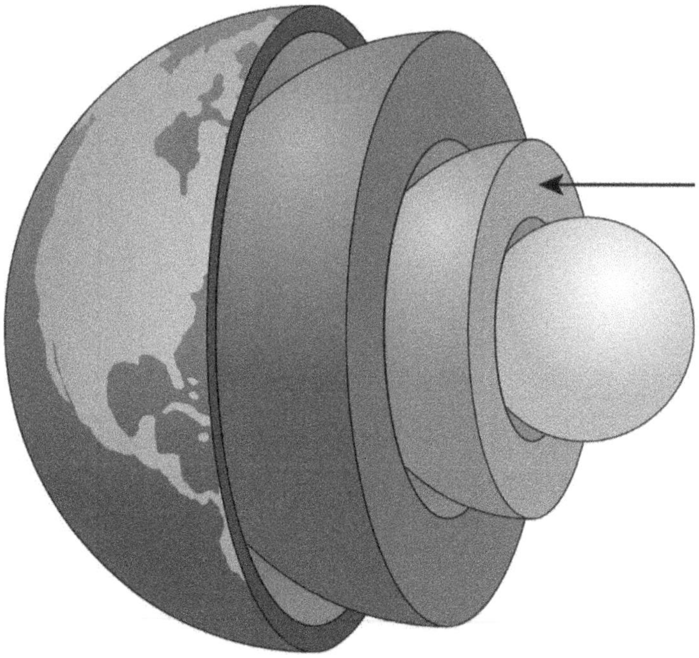

   (a) Name the layer indicated. Circle the correct answer.

   A    crust

   B    inner core

   C    mantle

   D    outer core

   [1]

   (b) Write down the layer of the Earth that includes tectonic plates.

   ....................................................

   [1]

**2** This question is about the Earth's atmosphere.

(a) The table describes components of the atmosphere on Earth.

| Component | Proportion of atmosphere | Natural processes |
|---|---|---|
| Nitrogen | Nearly $\frac{4}{5}$ | Not used in respiration or photosynthesis. |
| Oxygen | Nearly $\frac{1}{5}$ | Used up by respiration and produced by photosynthesis. |
| **X** | Around $\frac{1}{100}$ | Not used in respiration or photosynthesis. |
| **Y** | Around $\frac{4}{10\,000}$ | Produced by respiration and used up by photosynthesis. |
| Other gases | Less than $\frac{1}{1000}$ | Not used in respiration or photosynthesis. |

Name the missing components, **X** and **Y**.

**X** = ......................................................

**Y** = ......................................................

[1]

(b) Draw lines to match each **atmospheric pollutant** to its **effects**.

**Atmospheric pollutant**         **Effects**

| Carbon dioxide | • Causes acid rain that damages plants.<br>• Can cause breathing problems for people. |

| Smoke | • Causes air to look grey or brown.<br>• Over time, can cause cancer in people. |

| Sulfur dioxide | • Small amounts help warm the Earth to sustain life.<br>• Large amounts cause the Earth to warm up too much. |

[2]

**3** The diagrams show tectonic plate boundaries.

(a) Add **two** arrows to the diagram below to show how these plates move to cause an earthquake.

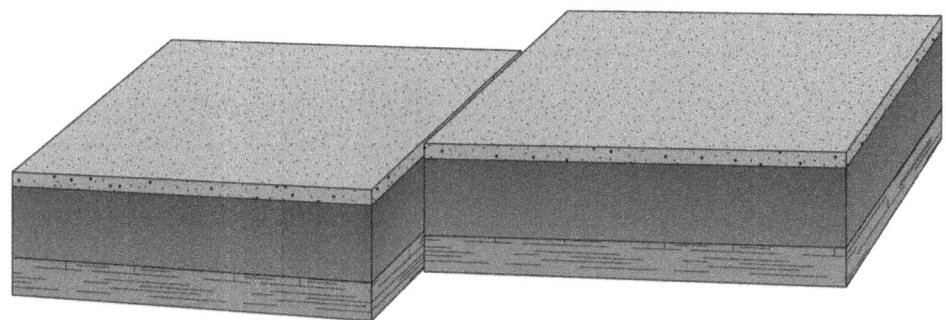

[1]

(b) Name the structure formed by the plate movement shown in the diagram below.

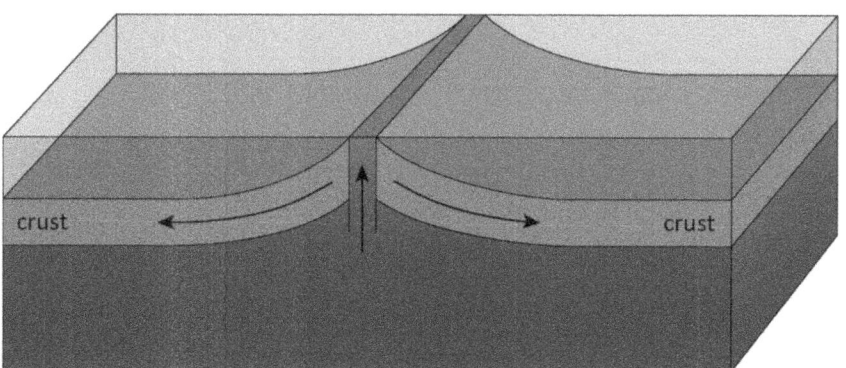

.......................................................
[1]

**4** Chen made a model of parts of the water cycle.

The diagram shows the apparatus.

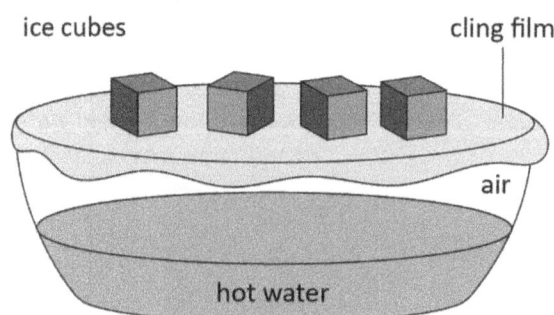

(a) Complete the sentences that describe two of the processes that Chen's model demonstrates.

The ice cubes cause water in the warm air to ................................... on the underside of the cling film.

Droplets of water that fall from the cling film represent ...................................
[1]

(b) There is a third process involving water that Chen cannot see, but which must be taking place.

Name this process.

.......................................................

[1]

(c) Think about all the processes that occur in the water cycle.

Describe a limitation (weakness) of Chen's model.

..................................................................................................................

..................................................................................................................

[1]

**5** The diagrams show how the surface of the Earth is thought to have changed over time.

(a) Explain how these diagrams support Wegener's hypothesis of continental drift.

..................................................................................................................

..................................................................................................................

[1]

(b) The continents of Africa and South America are moving apart at a rate of about 1.5 cm per year.

Calculate how far they move apart in 200 million years. Give your answer in kilometres.

.................................................. km

[1]

**6** Mia collects rainfall using the rain gauge shown in the diagram. She places it in the playground at her school.

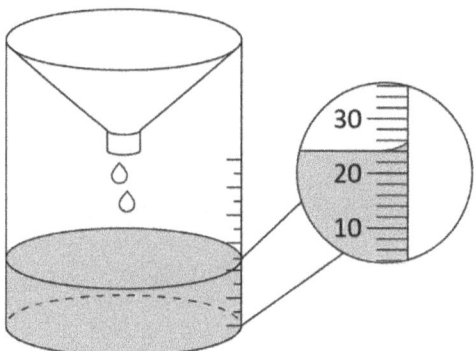

(a) Write down the water level in the rain gauge.

.................................................. mm
[1]

(b) Mia wants to measure the average rainfall around her village.

Suggest what other equipment she needs, if any, and explain your answer.

..................................................................................................................................

..................................................................................................................................
[1]

**7** Gabriella has a hypothesis that tectonic movement is causing mountains nearby to rise.

Gabriella proposes to take measurements to test this hypothesis.

(a) Do you think that Gabriella can test her hypothesis in this way? Explain your answer.

..................................................................................................................................

..................................................................................................................................
[1]

(b) Suggest other forms of evidence that Gabriella could look for.

..................................................................................................................................

..................................................................................................................................
[1]

**8** Hassan measures the composition of a sample of air.

The table shows his results.

| Component | Proportion of component in % |
|---|---|
| Carbon dioxide | 0.04 |
| Nitrogen | .............................. |
| Oxygen | 21 |
| Other gases | 0.96 |

(a) Use Hassan's results to calculate the proportion of nitrogen in the air.

Write the answer in the table.

[1]

(b) Use your result from (a) to complete the pie chart.

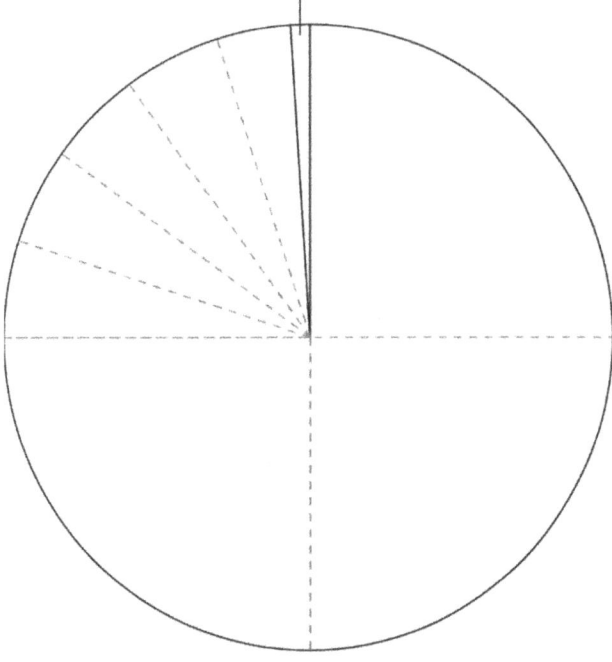

[1]

**9** Drinking water must be tested for purity and treated before being added to the water supply.

- Water has a low purity if it contains pollutants.
- It costs money to purify water.
- The more pollutants it contains, the more expensive it is to purify it.

Carlos tests two samples of water from each of four different sources.

The table shows his results.

| Source | Purity in % | | |
|---|---|---|---|
| | Sample 1 | Sample 2 | Mean |
| Rainfall | 97.5 | 95.5 | 96.5 |
| River | 89.0 | 96.0 | 92.5 |
| Well | 98.0 | 96.5 | 97.3 |
| Sea water | 85.5 | 88.5 | 87.0 |

(a) Which source would be the best supply of drinking water, based on Carlos's data? Explain why.

.................................................................................................................................

.................................................................................................................................
[1]

(b) The results collected by Carlos have low reliability. Explain what this means.

.................................................................................................................................

.................................................................................................................................
[1]

# Self-assessment and reflective learning page
# The Earth and its atmosphere

Tick (✓) the box that best shows how you feel about each statement.

| Statement | I don't know | I need more practice | I understand |
|---|---|---|---|
| I understand that the Earth's crust is divided into several tectonic plates that 'float' on the mantle. | | | |
| I can describe how two tectonic plates may move apart from each other or push against each other. | | | |
| I know that earthquakes and volcanic eruptions are sudden and violent events that can occur due to the forces between two plates at a plate boundary. | | | |
| I know that two plates colliding over millions of years can form fold mountains. | | | |
| I know that the Earth's atmosphere is a mixture of 78% nitrogen, 21% oxygen and smaller amounts of gases including carbon dioxide and argon. | | | |
| I understand that the mixture of gases in the Earth's atmosphere can change because of natural processes and human activities. | | | |
| I can describe the stages of the water cycle. | | | |
| **Thinking and working scientifically** | | | |
| I can describe the strengths and limitations of a model. | | | |
| I can identify whether a given hypothesis is testable. | | | |
| I can plan a range of investigations of different types. | | | |
| I can describe how to sort, group and classify data through testing and observation. | | | |
| I can decide what equipment is required to carry out an investigation and how to use it appropriately. | | | |
| I can evaluate whether measurements have been repeated sufficiently to be reliable. | | | |
| I can describe how to take appropriately accurate and precise measurements, explaining why accuracy and precision are important. | | | |
| I can describe how to collect and record data in an appropriate form. | | | |
| I can make conclusions by interpreting results. | | | |
| I can present and interpret observations and measurements appropriately. | | | |

Mark for end of unit test _____ /20 marks

| **What went well in this topic?** |
|---|
| |

| **What could you do better next time? What parts of the course could your teacher go through in a revision lesson that would support your improvement?** |
|---|
| |

| **Teacher comment:** |
|---|
| |

# End of Unit Test: The Earth in space
# Total = 20 marks

Name: .................................................  Class: ...................................................

Date: ....................................................

1  (a)  The statements describe how astronomers think the solar system formed.

Number the statements from 1 to 5 to show the correct order.

The first statement has been numbered for you.

| 1 | **A** The effects of gravity pulled parts of a huge cloud of dust and gas together. |

| | **B** As more gas was pulled inwards to the central 'ball', the particles in the gas squashed together. This caused nuclear reactions to start and formed the Sun. |

| | **C** The force of gravity squashed the particles of gas and dust in the smaller 'balls' together to form planets containing rocks, liquids and gases. |

| | **D** A 'ball' of gas and dust formed at the centre of the cloud and started to spin. The rest of the cloud formed a disc. |

| | **E** Parts of the disc of gas and dust pulled together to form smaller spinning 'balls'. |

[2]

(b)  Read the following statements, which describe observations of the solar system.

   A   Many planets have rocky moons in orbit around them.

   B   All planets receive thermal energy transferred by the Sun.

   C   Some planets are mostly made of rock, but other planets are mostly made of gas.

   D   Rocky planets and moons have craters caused by collisions with smaller pieces of rock.

   E   The Sun and planets have magnetic fields.

Which **two** statements provide direct evidence for the action of gravitational forces?

1. ................. and  2. .................

[2]

**2** Brianna lives near the sea.

Twice a day she observes that the water rises high up the beach.

In between these times, the water falls low down the beach.

(a) Name the phenomenon that Brianna has observed.

.................................................
[1]

(b) Write down the force responsible for this phenomenon.

.................................................
[1]

**3** The diagram shows an eclipse.

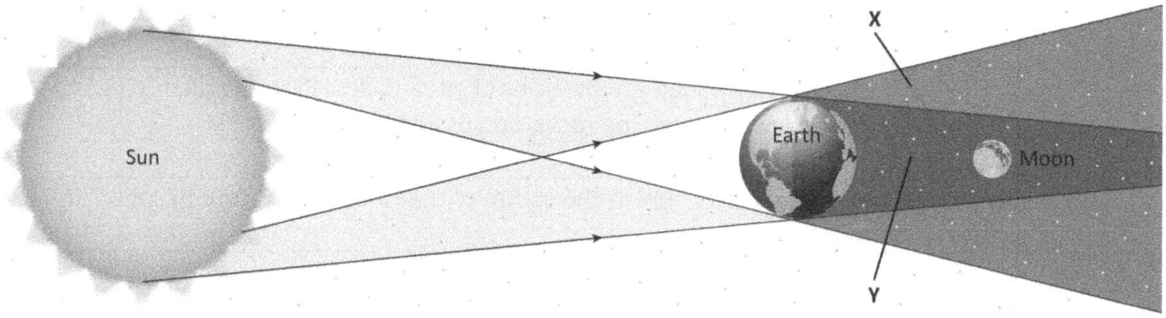

(a) Write down the type of eclipse.

............................................ ..................................................eclipse

[1]

(b) Name the regions labelled **X** and **Y**.

X = .................................................

Y = .................................................

[1]

4  Oliver hears from his teacher that there will be a solar eclipse soon.

   He predicts that the sky will get darker until the Moon appears to cover the Sun completely.

   He uses a telescope to observe the eclipse.

   He shines the light from a telescope onto a piece of paper.

   He does not look directly into the telescope.

   The image shows what he observes.

   (a) Explain why Oliver shone the light onto paper.

   ......................................................................................................................................

   ......................................................................................................................................
   [1]

   (b) Evaluate Oliver's prediction.

   ......................................................................................................................................

   ......................................................................................................................................
   [1]

**5** This question is about tides.

(a) The diagram shows the relative positions of the Earth, Moon and Sun at different times.

Which position of the Moon, A, B, C or D, will cause the greatest difference between low and high tide on Earth?

..................................................
[1]

(b) Safia measures the height of low tide and high tide on five different days.

The table shows her results.

| Day | Height in m | | Difference in m |
|---|---|---|---|
| | Low tide | High tide | |
| 1 | 1.5 | 4.9 | 3.4 |
| 2 | 1.1 | 5.2 | .......................... |
| 3 | 0.9 | 5.6 | .......................... |
| 4 | 0.6 | 5.8 | .......................... |
| 5 | 0.8 | 5.5 | .......................... |

(i) Calculate the differences between high and low tide on each day.

Write your answers in the table. One has been done for you.

[1]

(ii) Write down the day on which the difference between high and low tide was the greatest.

..................................................
[1]

**6** Read the following description of an experiment and answer the questions that follow.

In the solar system, there are many pieces of rock that are separate from the planets and their moons.

In 2023, a space probe called Osiris-Rex collected 0.25 kg of dust from the surface of one of these pieces of rock. The probe sealed the sample inside a tough, heat-proof vacuum container and returned it to Earth.

(a) Explain why the container is heat-proof.

.................................................................................................................................................

.................................................................................................................................................

[1]

(b) Scientists think that these rocks may be billions of years old. Suggest why collecting samples may provide evidence of how the solar system formed.

.................................................................................................................................................

.................................................................................................................................................

[1]

**7** The graph shows the distance between Earth and the planet Venus over several months.

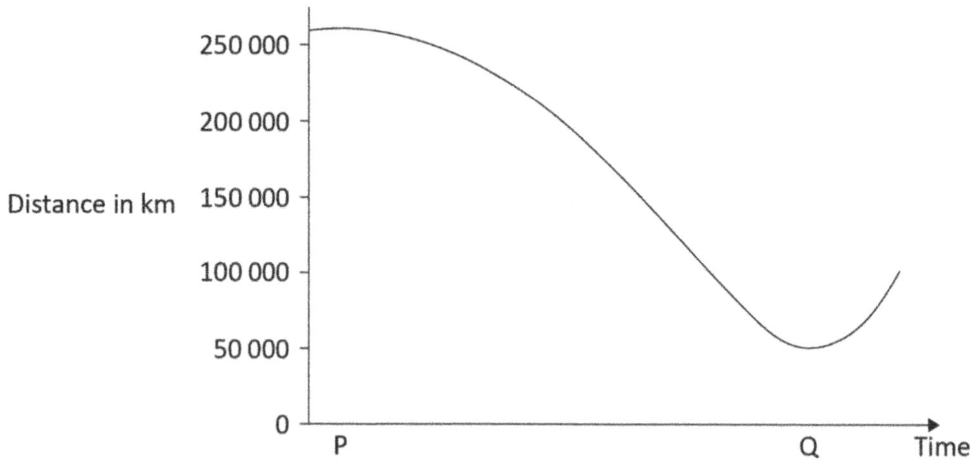

[Graph: Distance in km (y-axis, 0 to 250 000) vs Time (x-axis, with points P and Q marked). Curve starts at ~260 000 at P, decreases to minimum ~50 000 at Q, then rises slightly.]

(a) Complete the sentences to explain the graph.

Earth and Venus both ............................................. the Sun, but at different

............................................. from it. Sometimes they are on the ............................................. side

of the Sun, and sometimes they are on ............................................. sides.

[2]

(b) Suggest **one** way that the appearance of Venus changes from time **P** to time **Q**, when observed from Earth.

..................................................................................................................................................

[1]

**8** Priya needs help planning an investigation into tides.

She measures the height of sea water at different times for many days.

(a) Write down the dependent variable.

..................................................................................................................................................

[1]

(b) Priya wants to be sure that she observes at least **two** full cycles of spring tide and neap tide.

Suggest how much time is needed for Priya's investigation.

..................................................................................................................................................

..................................................................................................................................................

[1]

# Self-assessment and reflective learning page
# The Earth in space

Tick (✓) the box that best shows how you feel about each statement.

| Statement | I don't know | I need more practice | I understand |
|---|---|---|---|
| I know that the Sun and planets were formed from a solar nebula of gas and dust drawn together by the force of gravity. | | | |
| I know that the force of gravity holds planets in orbit around the Sun, and moons in orbit around planets. | | | |
| I understand that tides are caused by the effects on ocean water of the force of gravity due to the Moon and the Earth's spin. | | | |
| I know that a solar eclipse occurs when the Moon blocks the Sun's light from reaching the Earth's surface. | | | |
| I know that a lunar eclipse occurs when the Earth blocks the Sun's light from reaching the Moon's surface. | | | |
| **Thinking and working scientifically** | | | |
| I can identify whether a hypothesis is testable. | | | |
| I can predict likely outcomes for a scientific enquiry based on scientific knowledge and understanding. | | | |
| I can plan a range of investigations of different types, while considering variables appropriately. | | | |
| I can decide what equipment is required to carry out an investigation and how to use it appropriately. | | | |
| I can describe how to carry out practical work safely. | | | |
| I can evaluate a range of secondary information sources. | | | |
| I can describe the accuracy of predictions, based on results, and suggest why they were or were not accurate. | | | |
| I can describe trends and patterns in results. | | | |
| I can make conclusions by interpreting results and explain the limitations of the conclusions. | | | |
| I can present and interpret observations and measurements appropriately. | | | |

Mark for end of unit test _____ /20 marks

| **What went well in this topic?** |
|---|
|   |

| **What could you do better next time? What parts of the course could your teacher go through in a revision lesson that would support your improvement?** |
|---|
|   |

| **Teacher comment:** |
|---|
|   |

End of Year Test 1

Total = 50 marks

Name: .................................................  Class: ..................................................

Date: ..................................................

1 (a) Draw the arrangement of particles in a solid, a liquid and a gas.

Use circles to represent the particles in your diagram.

One circle has already been drawn in each box.

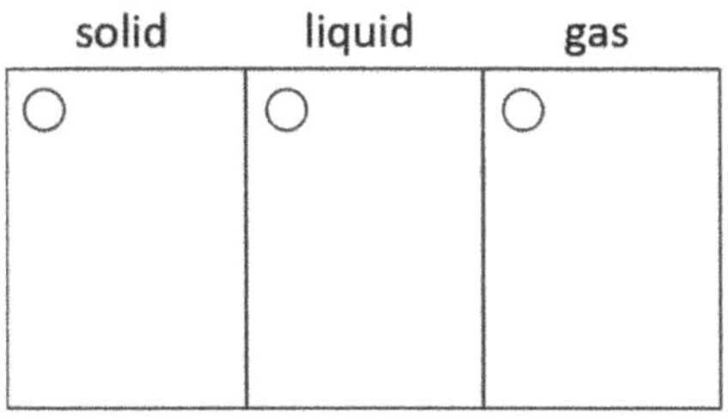

| solid | liquid | gas |
|---|---|---|
| ○ | ○ | ○ |

[3]

(b) Which of the following describes the changes that happens when a substance condenses?

Tick [✓] the correct box.

Solid to liquid  ☐

Gas to liquid  ☐

Solid to gas  ☐

Liquid to gas  ☐

[1]

(c) The properties of gases are different from those of liquids and solids.

Complete these sentences to describe these differences.

Particles in a gas will ..................................................... the container they are stored in.

Unlike liquids and solids, gases can be ..................................................... . This means that the volume of a gas can be decreased.

[2]

(d) Angelique wanted to investigate the particles in a gas. She wrote down a hypothesis:

| Particles of gas will get closer together if the gas is put into a freezer. |

(i) Explain why Angelique's hypothesis is **not** testable.

..................................................................................................................

..................................................................................................................
[1]

(ii) When a gas gets warmer, the particles move apart, and the balloon gets bigger.

Suggest how Angelique could use a balloon to investigate the effect of temperature on particles in a gas.

..................................................................................................................

..................................................................................................................

..................................................................................................................

..................................................................................................................
[2]

**2 (a)** Mould is involved in the decay of biological material. It plays an important ecological role.

(i) Write down the name given to organisms that decay biological material.

..................................................................................................................

..................................................................................................................
[1]

(ii) Describe the role of these organisms in ecosystems.

..................................................................................................................

..................................................................................................................
[1]

(b) Chen found some white mould growing on a piece of cake.

He wanted to investigate how quickly mould grows on the cake.

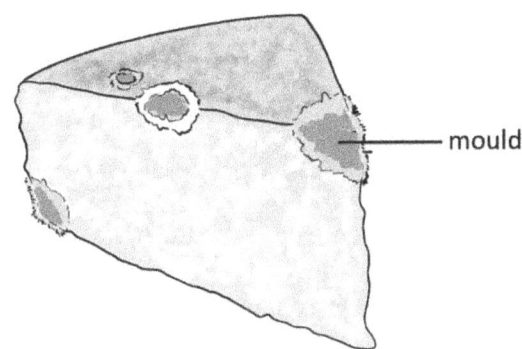

Chen used a piece of fresh cake and stored it in the open for 14 days.

He recorded his observations in a table.

These are Chen's results.

| Day | Percentage of the surface of cake covered in mould |
|---|---|
| 0 | 0 |
| 2 | 15 |
| 4 | 40 |
| 6 | 70 |
| 8 | 85 |
| 10 | 90 |
| 12 | 95 |
| 14 | 100 |

(i) Describe the pattern shown in Chen's results.

..................................................................................................................

..................................................................................................................
[1]

(ii) Chen's teacher says that the reliability of his results is low.

Explain how Chen could increase the reliability of his results.

..................................................................................................................

..................................................................................................................
[1]

**3** Most cells in plants and animals are specialised.

(a) The diagram shows two specialised cells from a plant.

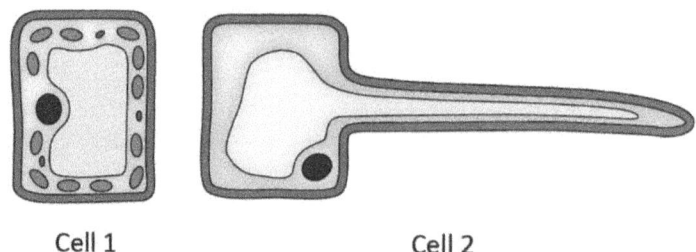

Cell 1           Cell 2

(i) Which cell is found in a leaf? Explain how you know.

.......................................................................................................................

.......................................................................................................................
[1]

(ii) Which cell is found in a plant root? Explain how you know.

.......................................................................................................................

.......................................................................................................................
[1]

(b) The diagram shows how a red blood cell is produced from an unspecialised cell in humans.

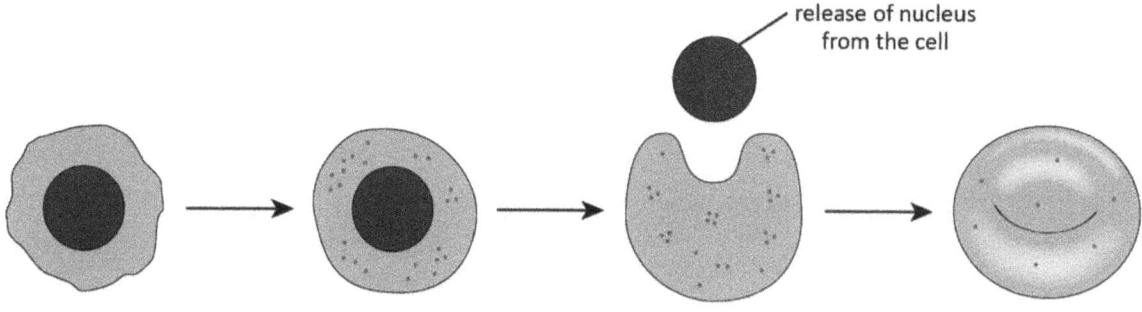

unspecialised cell                                                         mature red blood cell

(i) Red blood cells have a biconcave shape. This means that they have a dent in the middle of each side.

Write down why red blood cells have a biconcave shape.

.......................................................................................................................

.......................................................................................................................
[1]

(ii) The dots inside the cytoplasm of the developing cell represent haemoglobin.

Write down the function (job) of haemoglobin in a red blood cell.

................................................................................................................................

................................................................................................................................
[1]

(iii) Suggest why the nucleus is removed **after** haemoglobin has been formed.

................................................................................................................................

................................................................................................................................
[1]

**4** Different parts of a cell have different functions.

(a) Five parts of a cell are listed below.

cell membrane     cell wall     nucleus
cytoplasm     sap vacuole

(i) Complete the table to group the parts of a cell into those found in plants **only** and those found in animals **and** plants.

| Plants only | Animals and plants |
|---|---|
|  |  |
|  |  |
|  |  |

[2]

(ii) Draw lines to match each part of a cell to its function.

One has been done for you.

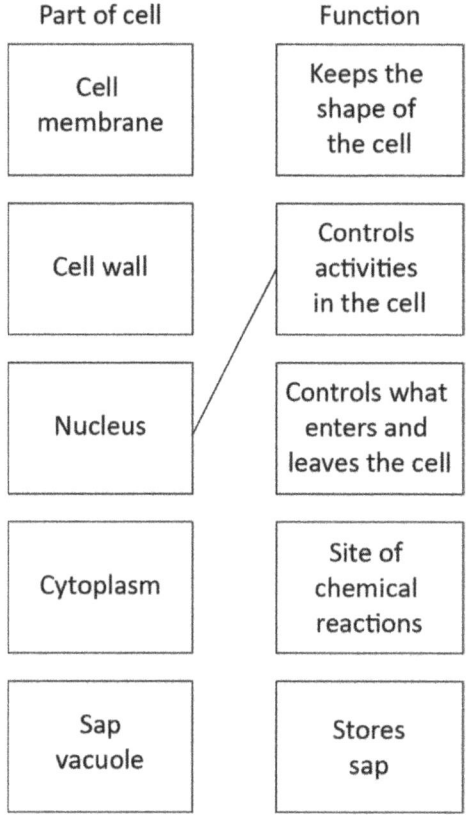

[2]

(b) Megumi is doing some research online into the features of a yeast cell.

Yeast cells have features of both animal and plant cells.

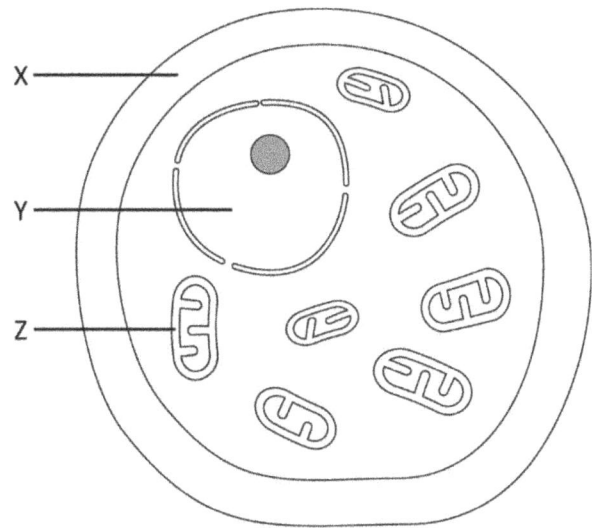

Megumi writes down that the yeast cell is more like a plant cell than an animal cell because it has structures labelled X, Y and Z.

Do you agree with Megumi?

Explain your answer.

...........................................................................................................................................

...........................................................................................................................................

...........................................................................................................................................

...........................................................................................................................................
[2]

**5 (a)** Write down a reason that scientists describe bacteria as microorganisms.

...........................................................................................................................................

...........................................................................................................................................
[1]

(b) The diagram shows five types of microorganism called bacteria, which all consist of at least one cell. They have been labelled with the letters V, W, X, Y and Z.

A scale bar is shown.

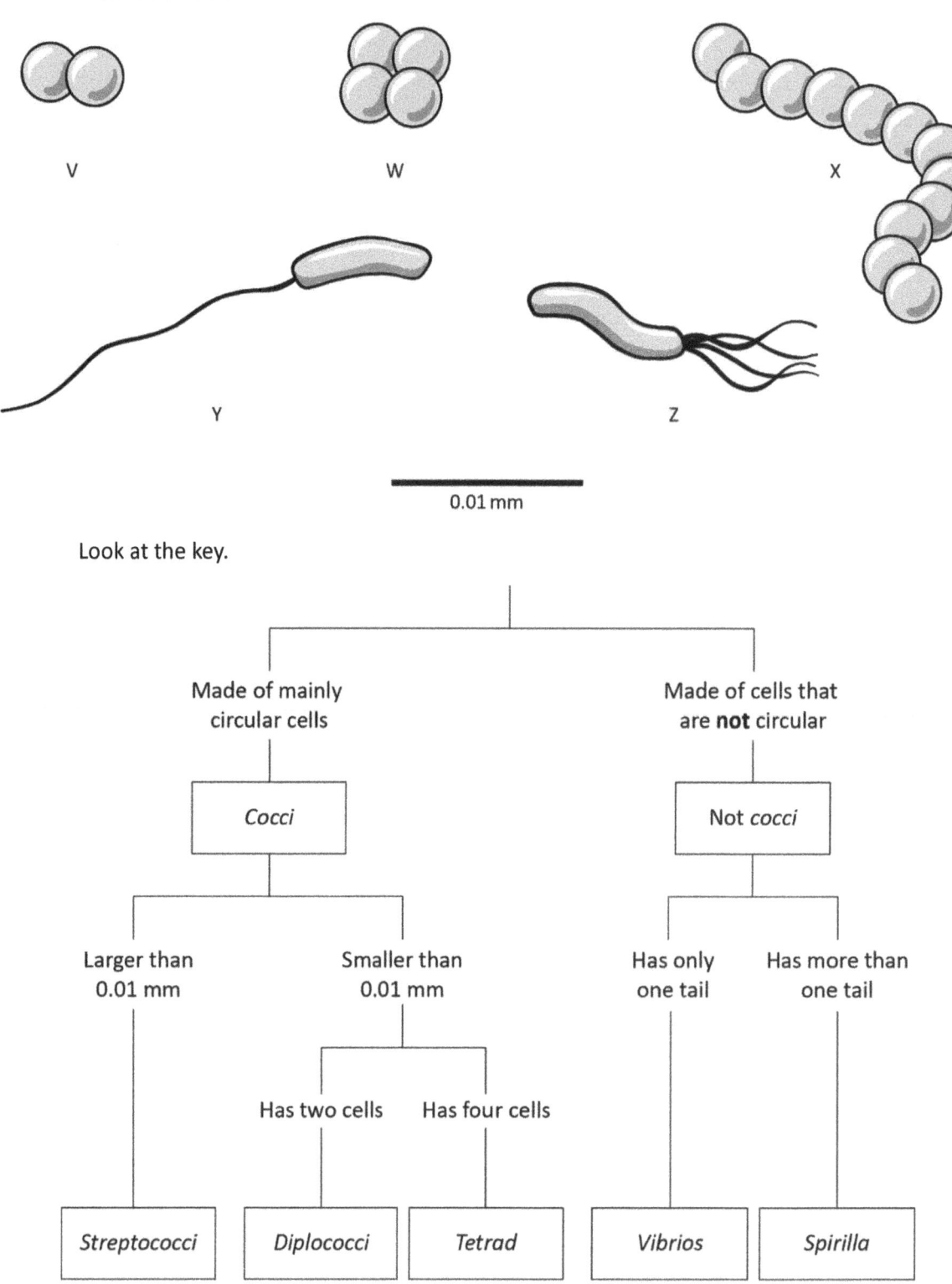

Use the key to find the names of the bacteria labelled V, W, X, Y and Z.

Write your answers in the table.

| name | letter |
|---|---|
| *Diplococci* | .................................................... |
| *Spirilla* | .................................................... |
| *Streptococci* | .................................................... |
| *Tetrad* | .................................................... |
| *Vibrios* | .................................................... |

[2]

**6** Metals and non-metals have different properties.

Haowen was provided with five solid objects by his teacher. They are listed below.

**plastic rod**  **iron nail**  **cling film**  **tin foil**

**copper wire**  **sheet of paper**  **glass cup**

(a) List the metals that were given to Haowen.

.................................................................................................................................................

[1]

(b) List the non-metals that were given to Haowen.

.................................................................................................................................................

.................................................................................................................................................

[1]

(c) Some metals and non-metals are liquids at room temperature (about 25 °C).

An example of a metal that is liquid at room temperature is mercury. An example of a non-metal that is liquid at room temperature is water.

Circle **two** physical properties that are likely to apply to mercury but not water at room temperature.

**hard**          **strong**          **malleable**

**good conductor of heat**          **brittle**

**good conductor of electricity**

[2]

**7** This question is about alloys.

(a) Complete the sentence about alloys.

Alloys are ................................... that have ................................... chemical and physical properties from their constituent substances.

[2]

(b) The table gives some information about two types of alloy.

| Alloy | Main elements found in this alloy |
|---|---|
| Duralumin | Aluminium and copper |
| Stainless steel | Iron and chromium |

Rajiv used sources on the internet to find out about the properties of duralumin and stainless steel.

He found that:

- duralumin is very lightweight and strong but corrodes (breaks down) quickly
- stainless steel is heavy and corrodes slowly.

(i) Rajiv suggested that duralumin would be a good alloy to build aircraft.

Do you agree with Rajiv?

..................................................................................................................

Explain your answer.

..................................................................................................................
[1]

(ii) Rajiv's friend suggested that stainless steel would be a better alloy to build aircraft.

Do you agree with Rajiv's friend?

..................................................................................................................

Explain your answer.

..................................................................................................................
[1]

(c) The elements found in the alloys in the table are mostly metals. However, alloys can contain non-metals.

Write down a **non**-metal element that is used in some types of alloy, including mild steel.

..................................................................................................................
[1]

**8** The drawings show six objects. Each of the objects transfers energy from one form to others.

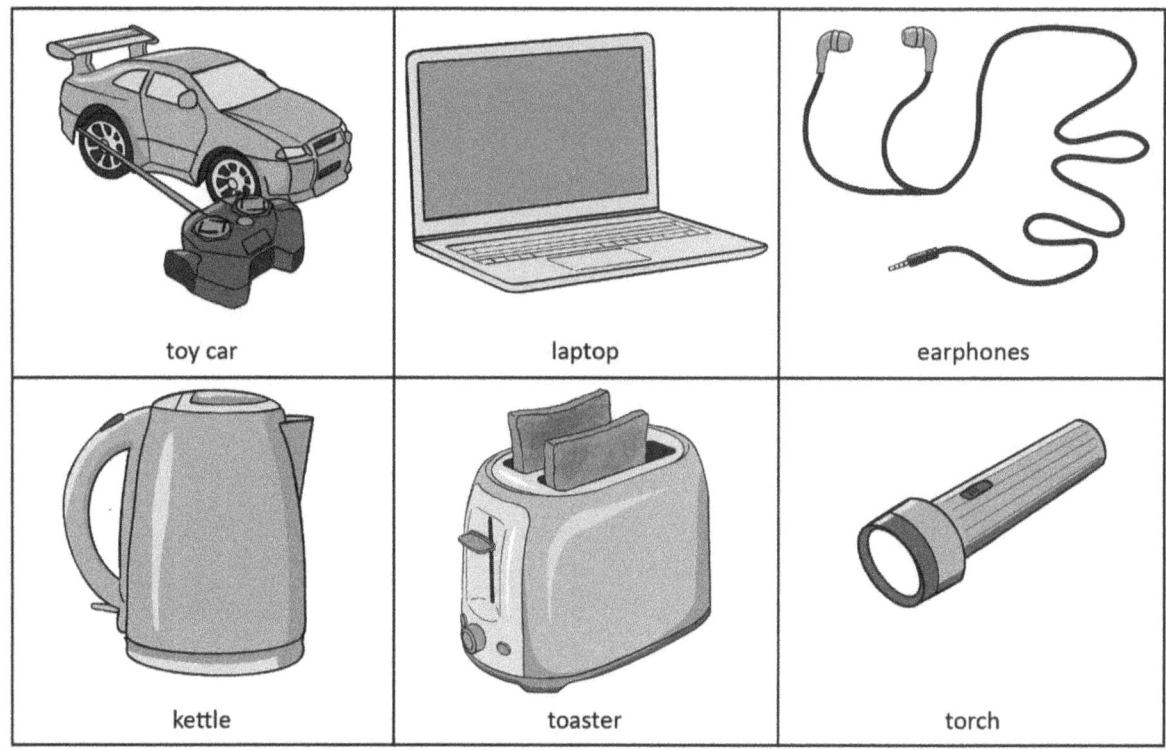

(a) Write down the names of **two** objects that transfer electrical energy into **useful** heat energy.

................................................................................................................................

................................................................................................................................
[2]

(b) Write down the name of one object that transfers energy into **wasteful** heat energy.

................................................................................................................................
[1]

(c) The loss of heat energy as waste energy is called energy dissipation.

Write down one **other** way in which energy is dissipated by the toy car.

................................................................................................................................
[1]

**9** In 1833, a scientist called Michael Faraday investigated the properties of a compound called silver sulfide. This contains atoms of the elements silver and sulfur.

(a) Explain why silver sulfide is called a compound and **not** a mixture.

.................................................................................................................................................

.................................................................................................................................................
[1]

(b) Faraday completed experiments and drew his results on a graph.

*Rate of electron flow through silver sulfide* vs *Temperature*

Explain why it is difficult from Faraday's results to decide whether silver sulfide is an electrical conductor or an electrical insulator.

.................................................................................................................................................

.................................................................................................................................................

.................................................................................................................................................

.................................................................................................................................................
[2]

(c) A student connected a sample of silver sulfide in a circuit, as shown in the diagram.

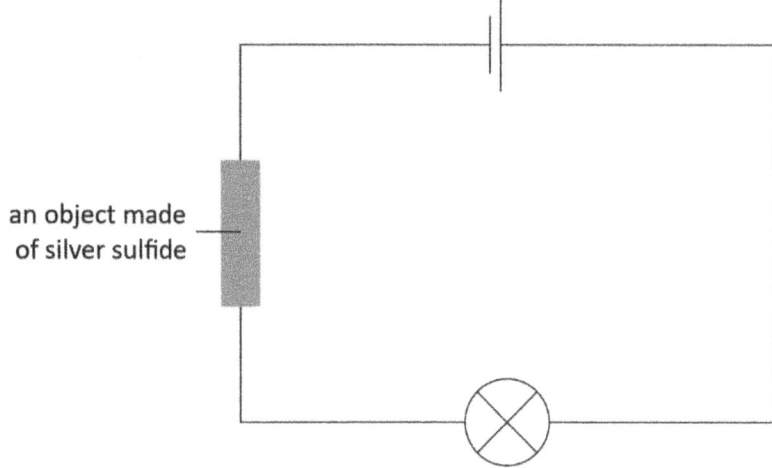

(i) Draw an arrowhead on a wire in the circuit to show the direction of electron flow in the circuit.

[1]

(ii) Sketch a line on the graph below to predict how the brightness of the lamp would change as temperature increases.

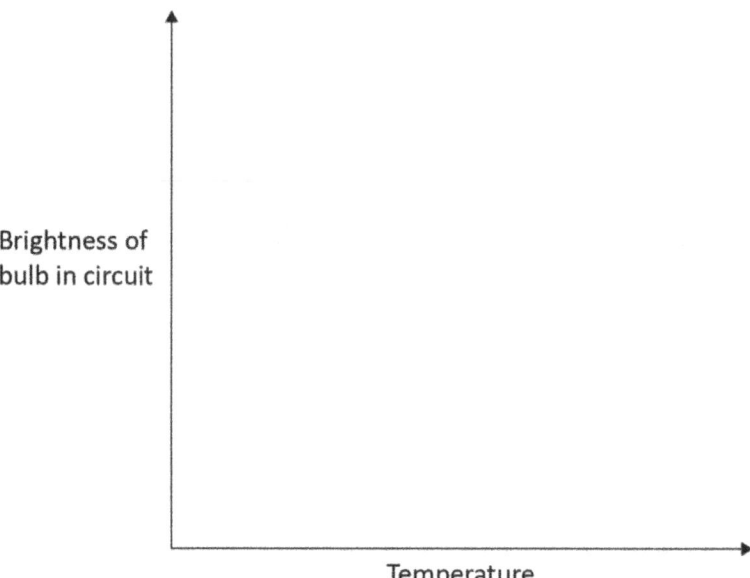

[1]

**10** Air is a mixture of gases.

(a) The pie chart shows the proportions of gases in clean, dry air. One gas is labelled X.

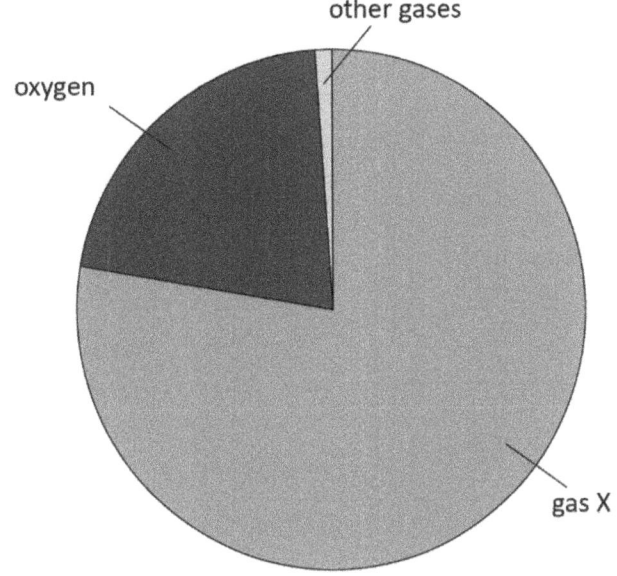

Identify gas X.

.................................................................................................................................................

[1]

(b) Air pollution is a serious concern.

(i) One type of air pollution is caused by cutting down trees and burning wood.

Scientists have made predictions about the proportions of gases in air in the year 2100 if these practices continue.

Suggest **two** of these differences.

1. ...................................................................................................................

2. ...................................................................................................................

[2]

(ii) Acid rain is another effect of pollution caused by humans.

Name **two** gases that are the cause of acid rain.

1. ...................................................................................................................

2. ...................................................................................................................

[2]

# End of Year Test 2
# Total = 50 marks

Name: ................................................. Class: ....................................................
Date: ...................................................

1  Look at the table of data about objects in the solar system.

| Object  | Mass relative to Earth |
|---------|------------------------|
| Mercury | 0.06                   |
| Earth   | 1.00                   |
| Jupiter | 320                    |
| Sun     | 330 000                |

(a) Complete the sentences.

Mercury, Earth and Jupiter are held in orbit around the Sun.

The force that causes this is called ......................................................

[1]

(b) The Sun has many times more mass than Jupiter.

Calculate how many more times, to the nearest whole number. Show your working.

..................................................

[2]

2  A fungus uses the energy found in dead organisms.

(a) What type of organism is the fungus? Tick (✓) one box.

Decomposer ☐

Predator ☐

Prey ☐

Producer ☐

[1]

(b) The diagram shows a food chain.

Write the word 'fungus' in a box on this diagram. Draw arrows to show where it gets its energy.

[2]

**3** Look at the diagram.

It shows a cross-section through part of the Earth.

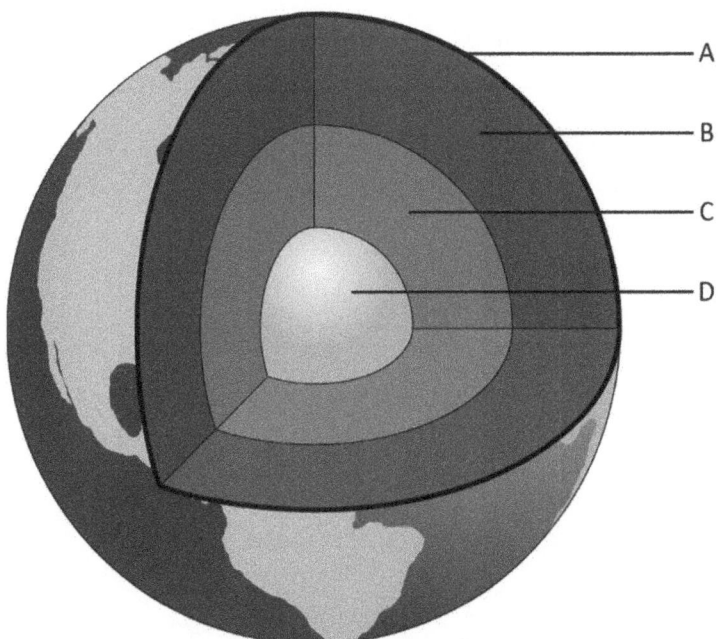

(a) Write down the letter that shows which layer, **A**, **B**, **C** or **D**, is the mantle.

....................................................

[1]

(b) Write down the letter that shows which layer, **A**, **B**, **C** or **D**, includes tectonic plates.

...................................................
[1]

(c) Explain how tectonic plates can move.

............................................................................................................

............................................................................................................
[1]

**4** A pathogen is something that causes a disease.

Many pathogens are viruses or bacteria.

(a) Give **one** reason why a virus is non-living.

............................................................................................................
[1]

The table shows some diseases caused by pathogens.

| Disease | Caused by pathogen |
|---|---|
| Chicken pox | Virus |
| COVID-19 | Virus |
| Measles | Virus |
| Tetanus | Bacterium |

(b) Living pathogens can be treated with antibiotics.

Which of the diseases in the table can be treated using antibiotics?

...................................................
[1]

5  The diagram shows the first 89 elements in the Periodic Table.

| 1 H | | | | | | | | | | | | | | | | | 2 He |
|---|---|---|---|---|---|---|---|---|---|---|---|---|---|---|---|---|---|
| 3 MLig | 4 Be | | | | | | | | | | | 5 B | 6 C | 7 N | 8 O | 9 F | 10 Ne |
| 11 Na | 12 Mg | | | | | | | | | | | 13 Al | 14 Si | 15 P | 16 S | 17 Cl | 18 Ar |
| 19 K | 20 Ca | 21 Sc | 22 Ti | 23 V | 24 Cr | 25 Mn | 26 Fe | 27 Co | 28 Ni | 29 Cu | 30 Zn | 31 Ga | 32 Ge | 33 As | 34 Se | 35 Br | 36 Kr |
| 37 Rb | 38 Sr | 39 Y | 40 Zr | 41 Nb | 42 Mo | 43 Tc | 44 Ru | 45 Rh | 46 Pd | 47 Ag | 48 Cd | 49 In | 50 Sn | 51 Sb | 52 Te | 53 I | 54 Xe |
| 55 Cs | 56 Ba | 57 La | 72 Hf | 73 Ta | 74 W | 75 Re | 76 Os | 77 Ir | 78 Pt | 79 Au | 80 Hg | 81 Tl | 82 Pb | 83 Bi | 84 Po | 85 At | 86 Rn |
| 87 Fr | 88 Ra | 89 Ac | | | | | | | | | | | | | | | |

(a) What does each box in the Periodic Table represent? Circle **one** answer.

**Compound**        **Element**        **Mixture**

[1]

(b) Based on your knowledge of the Periodic Table, state whether each of the highlighted boxes, **Cr** and **S**, is a metal or a non-metal.

**Cr**: ....................................

**S**: ....................................

[2]

(c) **Cr** and **S** are both solids at room temperature.

Roisin investigates whether **Cr** and **S** are hard or soft.

What kind of test should Roisin use? Tick (✓) **one** box.

Dissolving ☐

Melting ☐

Scratching ☐

Weighing ☐

[1]

**6** A ship measures the depth of the seabed using sound waves.

The ship emits a sound downwards through the water.

(a) Describe what happens to the sound wave when it reaches the seabed.

.................................................................................................................................................
[1]

(b) The diagram shows the ship.

Draw an arrow to complete the path of the sound wave.

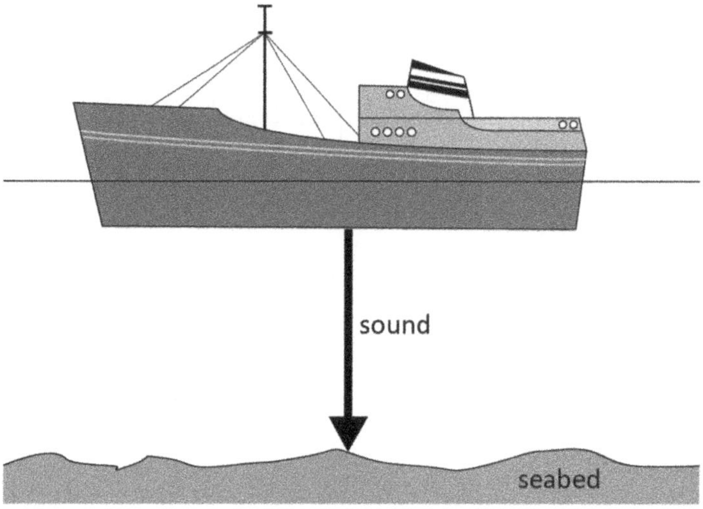

[1]

(c) The ship measures the total distance the sound wave travels as 500 m.

What is the depth of the seabed? Show your working.

.......................................... m
[2]

**7** Many cells in the human body are specialised.

The diagram shows three ciliated epithelial cells.

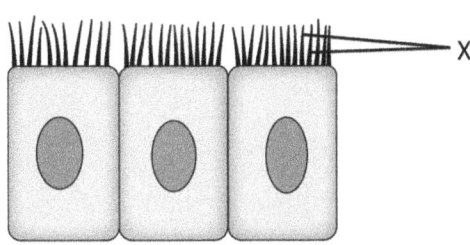

(a) Describe the function of the structures labelled **X**.

..................................................................................................................................................

[1]

(b) These cells are grouped together as part of the respiratory system.

Draw lines to match each **structure** with its **level of organisation** in the respiratory system.

| Structure | Level of organisation |
|---|---|
| Respiratory system | Tissue |
| Lung | Organ system |
| Group of ciliated cells | Organ |

[2]

**8** An appliance is connected to the electrical supply by a power cable.

(a) Circle the correct terms to complete the sentences:

The centre of the cable must be made of **a conductor / an insulator**.

The outer part of the cable should be made of **a conductor / an insulator**.

[2]

(b) Explain your choice for the outer part of the cable.

..................................................................................................................................................

..................................................................................................................................................

[1]

**9** Jinwoo measures the amounts of different gases in a sample of air.

| Gas | Percentage of sample |
|---|---|
| ............................. | 78% |
| Oxygen | 21% |
| Carbon dioxide | 0.05% |

(a) Complete the table by naming the gas in the first row.

[1]

(b) Jinwoo knows that there is one more gas in the sample, which is **not** shown in the table.

He has not been able to identify it.

Suggest the name of this gas.

..................................................................................................................................
[1]

(c) What percentage of the sample has **not** been identified? Show your working.

..................................................%
[2]

**10** Daniela measures the time it takes for a feather to fall inside a tube, from the top to the bottom.

Her investigation has two tubes.

Tube **1** is filled with air.

Tube **2** contains a vacuum.

(a) Explain the meaning of the term 'vacuum'.

..................................................................................................................................
[1]

(b) Which force on the feather is present in tube **1**, but **not** in tube **2**?

..................................................................................................................................
[1]

(c) Daniela repeats each part of the investigation three times.

Explain why she does this.

...................................................................................................................................

[1]

**11** Ewan investigates the reaction between an acid and an alkali.

He has some acid, some alkali, red litmus paper and blue litmus paper.

Ewan:

- measures an amount of alkali and puts it in a flask
- tests the solution with red litmus paper
- adds a small amount of acid
- then tests the solution again with red litmus paper
- keeps on adding the same small amount of acid and testing each time.

(a) What type of reaction is Ewan investigating?

Tick (✓) **one** box.

Corrosion ☐

Forming a precipitate ☐

Forming carbon dioxide ☐

Neutralisation ☐

[1]

(b) Explain how Ewan will know when the solution is no longer alkaline.

...................................................................................................................................

[1]

(c) Ewan checks the amount of acid added and thinks there is too much.

Describe what he can do to find out if the solution is now acidic.

...................................................................................................................................

[1]

(d) Ewan wants to make a solution that is pH 7.

He needs to change the equipment and method to do this more easily.

(i) Suggest what substance Ewan should use instead of litmus paper.

...................................................................................................................................
[1]

(ii) Describe the appearance of this substance when the solution has reached pH 7.

...................................................................................................................................
[1]

**12** Rubina investigates an electric circuit.

This is the equipment Rubina uses.

- Cells
- Lamps
- Switches
- Wires
- Ammeter

(a) Write down the function of the ammeter.

...................................................................................................................................
[1]

(b) The diagram below shows part of the circuit Rubina uses at the beginning of the experiment.

Draw the symbols for the **two** missing components to complete the diagram.

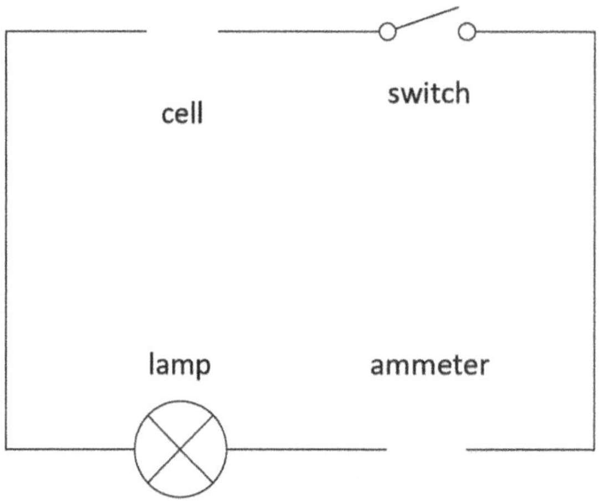

[2]

(c) Rubina wants to increase the current.

Which component should Rubina add to her circuit?

..................................................................................................................................................

[1]

**13** This list shows some things that living things can do.

| excrete | grow | move | reproduce |

(a) Choose the word, or words, from the list to complete the statement below.

A word can be used once, more than once or not at all.

A species is a group of organisms that can ..................................................... with one another

and have offspring that can also .....................................................

[1]

The pictures show two birds, species **A** and **B**.

(b) Use the following classification key to identify species **A** and **B**.

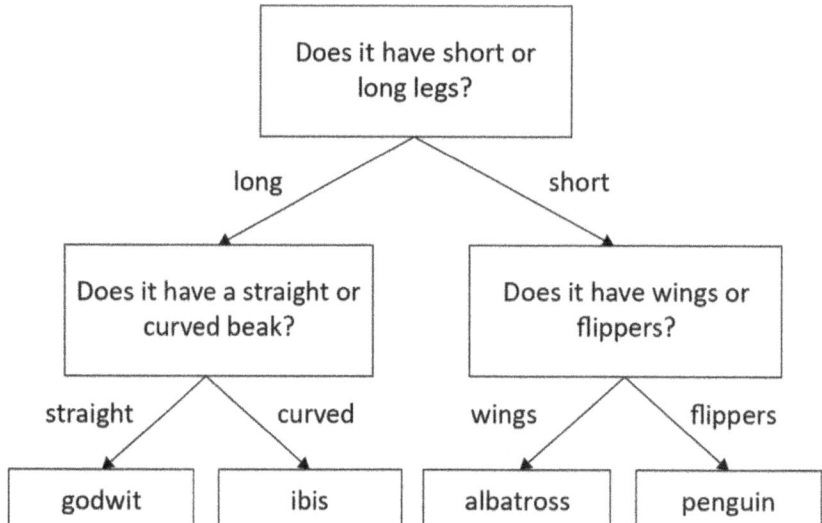

A: ..................................................

B: ..................................................

[2]

(c) Look carefully at the pictures. Suggest **one** other feature that you can see is different between birds **A** and **B**.

..........................................................................................................................................
[1]

**14** Skye tests a sample of a gas, using the equipment in the diagram.

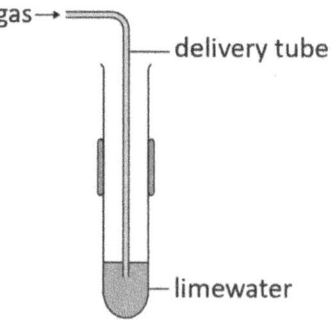

The limewater turns from colourless to milky white.

(a) Identify the gas.

..........................................................................................................................................
[1]

(b) This test is a chemical reaction between a gas and a solution.

The milky white substance is a solid.

What type of chemical reaction is this?

..................................................................................................................................................

[1]

(c) Suggest **two** safety precautions Skye should take when carrying out this test.

..................................................................................................................................................

..................................................................................................................................................

..................................................................................................................................................

..................................................................................................................................................

[2]

(d) Skye looks at the description of limewater in the hazards manual. This is what it says:

Moderate hazard of skin or eye irritation. Harmful if swallowed.

Which of the below is the most appropriate hazard symbol? Circle **one** symbol.

[1]

# Organisms and cells

| Qu | Answer | Mark | Further Information | Spec/TWS/AO |
|---|---|---|---|---|
| 1 | X – Cell membrane<br>Y – Nucleus | 2 | Each correct answer = 1 mark | 7Bs.02<br>AO1 |
| 2 | Mitochondria – Release energy through respiration<br>Cell membrane – Controls the movement of substances into and out of the cell | 2 | Each correct line = 1 mark | 7Bs.02<br>AO1 |
| 3(a) | Cell A – Ciliated cell<br>Cell B – Neurone | 2 | Also accept ciliated epithelial cell.<br>Each correct answer = 1 mark | 7Bs.03<br>AO1 |
| 3(b) | It has cilia/tiny hairs covered in mucus that trap dust/bacteria/debris<br>and move/sweep the particles. | 2 | One mark for each point (just saying that the ciliated cells has cilia/tiny hairs does not get a mark). | 7Bs.03<br>AO2 |
| 4(a) | Sap vacuole and cell wall | 2 | Also accept large vacuole.<br>One mark for each feature. | 7Bs.04<br>AO1/AO2 |
| 4(b) | It has a large surface area to increase the rate of absorption of water (and minerals). | 2 |  | 7Bs.03<br>AO1/AO2 |
| 5(a) | Move the stage (the flat ledge the slide sits on) down to its lowest position. = 1<br><br>Select the lowest power objective lens. = 2<br><br>Place the glass slide onto the stage. = 3<br><br>Turn the focusing wheel slowly until the cells are in focus and can be seen clearly. = 4 | 1 | All steps must be in correct sequence for 1 mark. | 7TWSc.05<br>7Bs.01<br>AO3 |
| 5(b) | 50/500 = 0.1 mm | 1 | Allow mark without unit. | 7TWSa.05<br>AO2 |
| 5(c) | Coverslip<br>Nuclear stain | 2 | One mark for each correct piece of equipment. | 7TWSm.02<br>AO2 |
| 6(a) | Cells | 1 |  | 7Bs.01<br>AO2 |

| Qu | Answer | Mark | Further Information | Spec/TWS/AO |
|---|---|---|---|---|
| 6(b) | Strength: It clearly shows the order of levels of organisation/relationship between the levels and helps show how smaller units of organisation such as cells, make up larger structures. Weakness: The model doesn't give any information about each level apart from the order. | 2 | Accept any other valid strengths or weaknesses. 1 mark for a strength. 1 mark for a weakness. | 7TWSm.01 6(b) AO3 |
| 6(c) | Organs | 1 | | 7Bs.01 AO2 |

# Microorganisms and classification

| Qu | Answer | Mark | Further Information | Spec/TWS/AO |
|---|---|---|---|---|
| 1(a) | Bird<br>Fungus | 2 | Each correct answer = 1 mark | 7Bp.02<br>AO1 |
| 1(b) | (i) Meal moth<br>(ii) Stick insect<br>(iii) Black kite | 1 | All three needed in correct boxes for 1 mark. | 7TWSc.01,<br>7Bp.04<br>AO2 |
| 1(c) | A group of organisms that can breed with one another to produce fertile offspring. | 1 | | 7Bp.03<br>AO1 |
| 2(a) | Nutrition | 1 | | 7Bp.01<br>AO1 |
| 2(b) | The release of energy from sugar. | 1 | Accept A chemical process (that happens inside cells) to release energy. | 7Bp.01<br>AO1 |
| 2(c) | Yeast – Single celled<br>Bacteria – Single celled<br>Insect – Multicellular | 2 | 3 lines drawn correctly for 2 marks.<br>2 lines drawn correctly for 1 mark.<br>1 or no lines drawn correctly for 0 marks. | 7Bs.01<br>AO1 |
| 2(d) | Because they do not complete all of the seven life processes. For example, viruses cannot move, reproduce, respond to their environment, use their own energy or excrete waste products. | 2 | 1 mark for stating viruses don't carry out all the seven life processes.<br>1 mark for giving at least one example of a life process they don't carry out. | 7Bp.01<br>7Bp.02<br>AO1 |
| 3(a) | 1 division every 20 minutes = 3 divisions in 60 minutes/1 hour.<br>2 hours × 3 divisions = 6 divisions in total.<br>Therefore, $1 \times 2^6 = 1 \times 64$ which is 64 bacteria. | 2 | 1 mark for calculating there would be 6 divisions.<br><br>1 mark for calculating the total to be 64. | 7TWSm.02<br>AO3 |
| 3(b) | Test tube B: Warm and moist | 1 | | 7TWSp.03<br>AO3 |
| 3(c) | To slow the growth and reproduction of microorganisms that cause the food to decompose/spoil. | 1 | | 7Be.01<br>AO1 |
| 4(a) | The type/batch of broth, the length of time he heated each flask, the size of flask or the temperature of flasks. | 1 | Award 1 mark for one correct variable. | 7TWSp.04<br>AO3 |

| Qu | Answer | Mark | Further Information | Spec/TWS/AO |
|---|---|---|---|---|
| 4(b)(i) | Microorganisms in the air got stuck in the curved neck. | 1 | | 7TWSa.03 AO3 |
| 4(b)(ii) | Microorganisms in the air were able to enter this flask. | 1 | | 7TWSa.03 AO3 |
| 4(c) | Yes, because only the broth in the flask with the straight neck spoiled. | 1 | | 7TWSa.01 AO3 |
| 5(a) | Yes | 1 | | 7TWSp.02 AO3 |
| 5(b) | Because only the jar that has meat exposed to the flies had maggots on the meat after two weeks. | 1 | | 7TWSp.02 AO3 |

# Structure and properties of materials 1

| Qu | Answer | Mark | Further Information | Spec/ TWS/AO |
|---|---|---|---|---|
| 1(a) | Particles arranged in a regular pattern and must be touching each other. | 1 | | 7Cm.06 AO1 |
| (b) | They/The particles vibrate. | 1 | Accept move up and down. | 7Cm.06 AO1 |
| (c) | Gas | 1 | Accept answer if it is indicated in another way, for example if the correct answer is underlined. Don't award the mark if more than one answer is circled. | 7Cm.06 AO1 |
| (d) | A space devoid/empty of matter. | 1 | Accept equivalent answers. Don't allow a space where there is only gas or air. | 7Cm.05 AO1 |
| 2 | B | 1 | Don't allow a mark if more than one box is ticked. | 7TWSp.05 AO1 |
| 3(a)(i) | A – red<br>B – alkali | 1<br>1 | Don't allow pink.<br>Allow any type of alkali, e.g. weak alkali. | 7Cp.02, 7Cp.03 AO2 |
| 3(a)(ii) | C – neutral<br>D – orange | 1<br>1 | Accept any spelling.<br>Accept red, yellow or any combination of red and orange or orange and yellow. | 7Cp.02, 7Cp.03 AO2 |
| 3(b) | Wear goggles/gloves to protect your eyes/skin. | 1<br><br>1 | Accept any reasonable suggestion.<br>The action should match the reason. | 7TWSc.05 AO3 |
| 3(c) | It tells you how acidic/or how alkaline the solution is/it tells you the pH of the solution. | 1 | Allow: It tells you if it is a weak/strong acid (or alkali). | 7Cp.03 AO1 |
| 4 | Acidity or alkalinity – Chemical property<br>Melting point – Physical property<br>Good electrical conductor – Physical property | 2 | All three correct answers required for 2 marks.<br>Allow any two correct answers for 1 mark.<br>Don't allow more than one line drawn from each description. | 7Cp.01, 7Cp.02 AO1 |
| 5(a)(i) | The bar for copper should go up to 1084. | 1 | Accept bars where the top of the bar is within 1 mm. The sides of the bar don't have to be drawn with a ruler. | 7TWSa.05 AO3 |
| 5(a)(ii) | Gold | 1 | Accept Au. | 7TWSa.05 AO3 |

| Qu | Answer | Mark | Further Information | Spec/TWS/AO |
|---|---|---|---|---|
| 5(b) | 1538 – 1064 = 474 | 1 | Accept 1538 – 1064 only, i.e. without an incorrect final answer. | 7TWSa.05 AO3 |
| 5(c) | To check the answer/peer review/check reliability. | 1 | Accept: To see if she has made a mistake/something has gone wrong. | 7TWSa.04 AO3 |
| 6(a) | Any one of:<br>It can explain the physical properties of solids/liquids/gases.<br>It can explain how matter changes state. | 1 | Accept an example such as: It explains why solids have a fixed shape. | 7TWSm.01 AO1 |
| 6(b) | Any one of:<br>Particles do have size/dimensions.<br>Particles do have mass.<br>There are forces between particles.<br>It doesn't take into account that there are different types of particle, e.g. atoms, molecules. | 1 | | 7TWSm.01 AO3 |

# Structure and properties of materials 2

| Qu | Answer | Mark | Further Information | Spec/TWS/AO |
|---|---|---|---|---|
| 1(a) | Atoms | 1 | Don't accept cells. | 7Cm.01 AO1 |
| 1(b) | Mixture | 1 | Accept other ways of indicating the answer, e.g. underlining. Don't award the mark if more than one answer is circled. | 7Cm.04, 7Cp.06 AO1 |
| 1(c) | It contains more than one type of atom/it contains atoms of two elements chemically joined together. | 1 | Allow: It contains carbon and oxygen atoms joined together. Don't allow: It contains two types of atom mixed together. | 7TWSm.02, 7Cm.04 AO1 |
| 1(d) | It does not have a fixed composition/the amount of the minerals can vary. | 1 | Don't allow: It is pure/natural. | 7Cm.04 AO2 |
| 2(a)(i) | Na | 1 | First letter must be a capital and second letter must be lower case. | 7Cm.04 AO2 |
| 2(a)(ii) | Iron | 1 | Allow any reasonable spelling. | 7Cm.04 AO2 |
| 2(b) | elements<br>metals/metal/metallic | 1<br>1 | Allow any reasonable spellings. Answers must be in the correct order. | 7Cm.03, 7Cm.02 AO1 |
| 3(a) | Good electrical conductor<br>Conducts heat well | 1<br>1 |  | 7Cp.05 AO1 |
| 3(b) | A way of heating all the metals equally for example placing each of the metals into a beaker of hot water.<br>The metal that heats up fastest is the best conductor. | 1<br>1 | Allow other ways to heat the metals equally.<br>Allow other feasible methods. | 7TWSp.04 AO3 |
| 4 | Pure metals have a regular/uniform structure so the layers can easily slip over each other.<br>Alloys have an irregular/non-uniform structure so the layers cannot easily slip over each other. | 1<br>1 |  | 7Cp.07 AO1 |
| 5(a) | C<br>F | 1 | Allow lower case letters. Allow letters in any order. Both letters required for the mark. | 7Cm.07, 7TWSm.02 AO2 |

| Qu | Answer | Mark | Further Information | Spec/TWS/AO |
|---|---|---|---|---|
| 5(b) | A<br>E | 1 | Allow lower case letters.<br>Allow letters in any order.<br>Both letters required for the mark. | 7Cm.07,<br>7TWSm.02<br>AO2 |
| 5(c) | B<br>D | 1 | Allow lower case letters.<br>Allow letters in any order.<br>Both letters required for the mark. | 7Cm.07,<br>7TWSm.02<br>AO2 |
| 6(a)(i) | Point plotted at 60% copper and 880°C. | 1 | Don't allow points drawn with a very thick pen or pencil. Do allow dots/points. | 7TWSc.07<br>AO3 |
| 6(a)(ii) | A straight line of best fit drawn through the points. | 1 | The line of best fit must extend at least as far as the first and last points.<br>Don't allow sketched lines. | 7TWSa.05<br>AO3 |
| 6(b) | As the percentage of copper in the alloy increases the melting point increases. | 1 | Accept the reverse argument.<br>Accept: It has a positive correlation. | 7TWSa.02<br>AO3 |

# Chemical changes and reactions

| Qu | Answer | Mark | Further Information | Spec/ TWS/AO |
|---|---|---|---|---|
| 1 | Chemical changes – Changing colour Temperature rise Fizzing Physical changes – Temperature remains constant A liquid turns into a gas Water freezes | 2 | Award 2 marks for all six correct. Award 1 mark for four or five correct. | 7Cc.01 AO1 |
| 2(a)(i) | Hydrogen – Lit splint – squeaky pop Oxygen – Relights a glowing splint Carbon dioxide – Turns limewater cloudy | 2 | All three lines correct for 2 marks. Two lines correct for 1 mark. Do not allow more than one line drawn from each gas. | 7Cp.04 AO1 |
| 2(b) | Flammable | 1 | Accept other ways of marking the box, for example crosses. Don't award a mark if more than one box is ticked. | 7TWSp.05 AO1 |
| 3(a) | Compounds | 1 | Don't award a mark if more than one box is ticked. | 7Cm.07 AO2 |
| 3(b) | Diagrams need to have only one type of particle shown by shading or labels. The particles must be touching and in a regular arrangement. | 1 | | 7Cm.01, 7Cm.07 AO2 |
| 4(a) | Precipitation | 1 | Accept any reasonable spelling. | 7Cc.02 AO2 |
| 4(b) | lead nitrate + potassium iodide → lead iodide + potassium nitrate | 1 | Accept reactants in any order. Accept products in any order. | 7TWSm.02 AO2 |
| 4(c) | Lead iodide is insoluble. | 1 | Accept: Lead iodide does not dissolve. | 7Cc.02 AO1 |
| 5(a) | Wear safety goggles/tie back long hair/do not touch hot equipment to protect your eyes/to stop your hair going in the flame/as it will burn you. | 1 | The explanation should match the safety precaution. Accept: Use a fume cupboard or mineral wool in the top of the test-tube so you do not breathe in sulfur dioxide. | 7TWSc.05 AO1 |
| 5(b) | Iron sulfide | 1 | Accept any reasonable spellings. Do not accept iron sulfate. | 7Cc.01 AO2 |

| Qu | Answer | Mark | Further Information | Spec/TWS/AO |
|---|---|---|---|---|
| 5(c) | The particles are rearranged and joined together. | 1 | Don't accept: There are more particles. | 7Cc.03 AO2 |
| 6(a)(i) | 7/seven | 1 | | 7Cc.04 AO1 |
| 6(a)(ii) | Neutralisation | 1 | | 7Cc.04 AO1 |
| 6(b) | A table with two headings: the label/name of the acid; the number of drops of alkali added. | 1 | | 7TWSc.07 AO3 |
| | The table should show A needed 30 drops, B needed 22 drops and C needed 15 drops. | 1 | | |
| 6(c) | Twice as much/2 times as much is added to test tube A compared to test tube C. | 2 | Award 1 mark for 15 drops more. | 7TWSa.05 AO3 |
| 6(d) | Same temperature
Same size of drop
Same concentration of alkali | 1 | Accept any reasonable answers. | 7TWSp.04 AO3 |

# Energy and forces

| Qu | Answer | Mark | Further Information | Spec/ TWS/AO |
|---|---|---|---|---|
| 1(a) | ↓ weight | 1 | Labelled arrow pointing vertically downwards = 1 mark Accept if arrow tail is from bottom edge of ball. Accept label: (force due to) gravity/force. | 7Pf.03, 7TWSm.02 AO1 |
| 1(b) | The ball moves/accelerates downwards/towards the ground. | 1 | | 7Pf.03 AO1 |
| 2(a) | Energy transfer from electrical components to internal energy store of water is **useful**. Energy transfer from electrical components to outer case of kettle and air is **wasted**. | 2 | Each correct word = 1 mark | 7Pf.02 AO1 |
| 2(b) | Battery transfers energy – by electric current to torch bulb. Candle transfers energy – by light to surroundings. Drum transfers energy – by sound to a person's ears. | 2 | All three correct = 2 marks One or two correct = 1 mark | 7Pf.01 AO1 |
| 3(a) | B, because there is no air in the tube/there is a vacuum in the tube. | 1 1 | Must include short explanation. | 7Pf.03, 7Pf.04, 7TWSc.02 AO1/AO3 |
| 3(b) | Air resistance | 1 | | 7Pf.04 AO1 |
| 4(a) | Dissipated | 1 | | 7Pf.01, 7Pf.02 AO1 |
| 4(b) | Temperature increases/it warms up | 1 | | 7Pf.01, 7Pf.02 AO1 |
| 5(a) | Desk lamp | 1 | | 7Pf.01, 7Pf.02, 7TWSa.02 AO2 |
| 5(b) | 50% | 1 | | 7Pf.01, 7Pf.02, 7TWSa.03 AO2 |
| 6(a) | Sun | 1 | | 7Pf.03 AO2 |

| Qu | Answer | Mark | Further Information | Spec/ TWS/AO |
|---|---|---|---|---|
| 6(b) | (Force due to) gravity depends on mass.<br>The Sun has the largest mass. | 1<br>1 | | 7Pf.03<br>AO2 |
| 7(a) | The more the elastic is stretched, the further the ball will travel/the distance increases. | 1 | | 7Pf.01,<br>7TWSp.03<br>AO2 |
| 7(b) | No (not safe), because (any one from):<br>• elastic may break and injure someone<br>• released ball may travel fast and hit someone<br>• direction of released ball is unpredictable/may cause damage. | 1 | Accept qualified 'yes (it is safe)' or 'can be made safe' only if a precaution is identified to avoid one of the safety issues described. Accept any other sensible safety issue. | 7Pf.01,<br>7TWSc.05<br>AO3 |
| 8(a) | Mass balance | 1 | Allow any other piece of apparatus that measures mass. | 7Pf.03,<br>7TWSc.02<br>AO3 |
| 8(b) | Any one from:<br>• Repeat steps 1 to 3 (at least 3 times) for each mass.<br>• Calculate the average (for each mass). | 1 | | 7Pf.03,<br>7Pf.04,<br>7TWSa.04<br>AO3 |

# Electricity and sound

| Qu | Answer | Mark | Further Information | Spec/TWS/AO |
|---|---|---|---|---|
| 1 | B | 1 | | 7Ps.01 AO1 |
| 2(a) | Lamp | 1 | | 7Pe.05, 7TWSm.02 AO1 |
| 2(b) | Battery (of cells) | 1 | | 7Pe.05, 7TWSm.02 AO1 |
| 3(a) | An electrical conductor is a material that **allows current to flow**. | 1 | | 7Pe.02 AO1 |
| 3(b) | Copper is an electrical **conductor**. Plastic is an electrical **insulator**, so it protects people from electric shocks. | 2 | Each correct term = 1 mark | 7Pe.02 AO1 |
| 4(a) | Echo, sound reflects on cliff | 1 | | 7Ps.02 AO1 |
| 4(b) | First peak shows original sound. Second peak shows echo. | 1 | Both points needed for 1 mark. Pointing out reduced loudness for echo is a valid point but does not score more marks. | 7Ps.02, 7TWSa.05 AO2 |
| 5(a) | Three arrows, one on each wire, pointing anticlockwise. | 1 | Accept if all arrows point in clockwise direction (students will not know about conventional current vs electron flow). 0 marks if one arrow is pointing in the opposite direction to the others. | 7Pe.01 AO1 |
| 5(b) | Flow of electrons stops/becomes zero. | 1 | | 7Pe.01, 7TWSp.03 AO2 |
| 6(a) | 1.2 A | 1 | Must include unit for 1 mark. Accept 1.20 A. | 7Pe.03, 7Pe.05, 7TWSc.04, 7TWSc.07 AO1 |
| 6(b) | Current will reduce. | 1 | Accept 'halve'. | 7Pe.04, 7TWSp.03 AO2 |

| Qu | Answer | Mark | Further Information | Spec/TWS/AO |
|---|---|---|---|---|
| 7(a) | Incorrect because the buzzer does not slow down or stop. | 1 | | 7Ps.01, 7TWSa.03 AO3 |
| 7(b) | Ideas including:<br>• sound requires a medium (to travel through)<br>• fewer air particles means less medium<br>• no particles/a vacuum means no medium, so sound cannot travel. | 1 | | 7Ps.01, 7TWSa.03 AO3 |
| 8(a) | Ammeter | 1 | | 7Pe.03, 7Pe.05, 7TWSm.02, 7TWSc.02 AO1 |
| 8(b) | Current increases as number of cells increases. | 1 | Note that the current is not linear/directly proportional to the number of cells. | 7Pe.04, 7TWSa.02 AO2 |
| 8(c) | Lamp breaks, so the circuit is broken and charge cannot flow. | 1 | Must include 'circuit broken' or similar wording.<br>Accept 'current too high' or 'lamp overheats' instead of 'lamp breaks'.<br>Accept 'current cannot flow' or 'no current present' instead of 'charge cannot flow'. | 7Pe.01, 7Pe.04, 7TWSa.05 AO3 |
| 9(a) | Loudness **50** dB   Reading **2** | 1 | No mark for stating only 50 dB or 'Reading 2'.<br>Accept 0.91 in place of 'Reading 2'. | 7Pe.03, 7TWSa.02 AO2 |
| 9(b) | 0.76 A | 1 | | 7Pe.03, 7TWSa.05 AO2 |
| 9(c) | $\left(\dfrac{0.92 + 0.92 + 0.95}{3}\right) = 0.93 \text{A}$ | 1 | Award 1 mark for answer stated with no working. | 7Pe.03, 7TWSa.05 AO3 |

# The Earth and its atmosphere

| Qu | Answer | Mark | Further Information | Spec/TWS/AO |
|---|---|---|---|---|
| 1(a) | Outer core | 1 | | 7ESp.01 AO1 |
| 1(b) | Crust | 1 | | 7ESp.01 AO1 |
| 2(a) | **X** = Argon, **Y** = Carbon dioxide | 1 | 1 mark for both components. 0 marks for one component. | 7ESp.03 AO1 |
| 2(b) | Carbon dioxide:<br>• Small amounts help warm the Earth to sustain life.<br>• Large amounts cause the Earth to warm up too much.<br><br>Smoke:<br>• Causes air to look grey or brown.<br>• Over time, can cause cancer in people.<br><br>Sulfur dioxide:<br>• Causes acid rain that damages plants.<br>• Can cause breathing problems for people. | 2 | All three matched for 2 marks. One or two correctly matched = 1 mark | 7ESp.03 AO1 |
| 3(a) | *[Diagram of two plates with arrows pointing in opposite directions parallel to plate boundary]* | 1 | Arrows must point parallel or nearly parallel to plate boundary. Arrows should point in opposite directions. Accept if **both** arrows are reversed compared to the answer given here. | 7ESp.02 AO1 |
| 3(b) | Volcano | 1 | Accept: volcanic eruption. | 7ESp.02 AO1 |
| 4(a) | The ice cubes cause water in the warm air to **condense** on the underside of the cling film. Droplets of water that fall from the cling film represent **precipitation**. | 1 | Both terms needed for 1 mark. 0 marks if only one term given correctly. Accept: 'condensation' for 'condense'. | 7ESc.01, 7TWSa.03 AO1 |
| 4(b) | Evaporation | 1 | | 7ESc.01, 7TWSa.03 AO1 |

| Qu | Answer | Mark | Further Information | Spec/TWS/AO |
|---|---|---|---|---|
| 4(c) | Any one from:<br>• cannot model other forms of precipitation (snow/sleet/hail, etc.)<br>• does not include other processes/parts of the water cycle (groundwater/run-off/rivers/lakes/clouds/glaciers/sea ice)<br>• idea that temperatures involved in the model are different from 'reality.' | 1 | Any named alternative form of precipitation.<br>Any named alternative process or part.<br>Accept any other reasonable limitation. | 7ESc.01,<br>7TWSm.01<br>AO3 |
| 5(a) | Ideas from:<br>• (tectonic) plates move<br>• continents/land on different plates move apart. | 1 | | 7ESp.01<br>AO1 |
| 5(b) | $\dfrac{1.5 \times 200\,000\,000}{100\,000} = 3000\,\text{km}$ | 1 | Working not required for full mark. | 7ESp.01<br>AO2 |
| 6(a) | 24 mm | 1 | Allow 23 or 25 but not less than 23 or more than 25. | 7ESc.01,<br>7TWSc.07<br>AO3 |
| 6(b) | Several more rain gauges (identical to the one she already has) to take more readings (and calculate an average). | 1 | Accept: 'more gauges placed around the village', although placement is not required. | 7ESc.01,<br>7TWSc.02<br>AO3 |
| 7(a) | No, because any one of:<br>• it takes many thousands of years/a very long time for mountains to rise noticeably<br>• measuring the height of a mountain is very difficult/needs complicated instruments. | 1 | | 7ESp.02,<br>7TWSp.01<br>AO2 |
| 7(b) | Ideas from:<br>• fossils on the mountain that would once have been on the seabed<br>• secondary evidence, e.g. historical measurements<br>• modern measurements, e.g. GPS satellite data. | 1 | Accept any other sensible idea. | 7ESp.02,<br>7TWSp.04<br>AO3 |

| Qu | Answer | Mark | Further Information | Spec/ TWS/AO |
|---|---|---|---|---|
| 8(a) | (100 − 21 − 0.96 − 0.04) = 78(%) | 1 | Only accept 78. | 7ESp.03, 7TWSa.05 AO2 |
| 8(b) | *Pie chart showing carbon dioxide and other gases (small segment), oxygen, and nitrogen (largest segment).* | 1 | Must include labels for mark. Accept nitrogen segment between 75 and 80% and oxygen segment between 20 and 24%. Allow other segment size if consistent with ECF from (a). | 7ESp.03, 7TWSa.05 AO2 |
| 9(a) | Well because highest purity/cheapest to treat/needs least treatment. | 1 | Choice and explanation needed for full mark. | 7ESc.01, 7TWSc.01 AO2 |
| 9(b) | Not very reliable with one reason from:<br>• large differences between some samples from the same source<br>• not enough samples tested. | 1 | | 7ESc.01, 7TWSc.03 AO2 |

# The Earth in space

| Qu | Answer | Mark | Further Information | Spec/ TWS/AO |
|---|---|---|---|---|
| 1(a) | 2 – **D**; 3 – **B**; 4 – **E**; 5 – **C** | 2 | 2 marks for all correct. 1 mark for 2 or 3 correct. 0 marks for 0 or 1 correct. | 7ESs.01 AO1 |
| 1(b) | **A** and **D** | 2 | 1 mark for each correct letter. | 7ESs.02, 7TWSc.06 AO1 |
| 2(a) | Tides | 1 | Accept: 'tide' or 'tidal movement'. | 7ESs.03 AO1 |
| 2(b) | Gravity | 1 | Accept: 'weight' or 'gravitational (force)'. | 7ESs.03 AO1 |
| 3(a) | (i) total (ii) lunar | 1 | **Both** words needed for 1 mark. | 7ESs.04 AO1 |
| 3(b) | **X** = penumbra **Y** = umbra | 1 | **Both** words needed for 1 mark. | 7ESs.04 AO1 |
| 4(a) | Any one from: <ul><li>it is dangerous to look directly at the Sun</li><li>reflecting the sunlight off (matte) paper reduces the brightness.</li></ul> | 1 | Accept: 'to protect his eyes' or similar description. | 7ESs.04, 7TWSc.05 AO1 |
| 4(b) | The prediction described a total (solar) eclipse, but a annular (solar) eclipse was observed. | 1 | Also accept a comment that indicates the prediction was incorrect, as long as it also mentions 'annular'. | 7ESs.04, 7TWSa.01 AO3 |
| 5(a) | B | 1 |  | 7ESs.03 AO2 |
| 5(b)(i) | <table><tr><th>Day</th><th>Difference in m</th></tr><tr><td>1</td><td>3.4</td></tr><tr><td>2</td><td>4.1</td></tr><tr><td>3</td><td>4.7</td></tr><tr><td>4</td><td>5.2</td></tr><tr><td>5</td><td>4.7</td></tr></table> | 1 | 1 mark for three or more values correct from days 2–5. | 7ESs.03, 7TWSa.05 AO2 |
| 5(b)(ii) | Day 4 | 1 |  | 7ESs.03, 7TWSa.02 AO2 |
| 6(a) | To survive the heat of re-entry to the atmosphere. | 1 | Accept: 'to protect the sample'. Accept other reasonable explanations. | 7ESs.01, 7TWSc.02 AO1 |

| Qu | Answer | Mark | Further Information | Spec/ TWS/AO |
|---|---|---|---|---|
| 6(b) | Ideas from:<br>• sample should be unchanged since the formation of the solar system<br>• identify the substances (in the sample) and compare with scientists' predictions<br>• determine the age (of the sample) and compare with scientists' predictions. | 1 | | 7ESs.01, 7TWSp.02 AO3 |
| 7(a) | Earth and Venus both **orbit** the Sun, but at different **distances** from it. Sometimes they are on the **same** side of the Sun, and sometimes they are on **opposite** sides. | 2 | 1 mark for both words correct in sentence 1.<br>1 mark for both words correct in sentence 2. | 7ESs.02, 7TWSa.03 AO2 |
| 7(b) | Any **one** from:<br>• gets brighter<br>• appears larger. | 1 | Also accept: 'changes phase'. | 7ESs.02, 7TWSp.03 AO2 |
| 8(a) | Height (of water) | 1 | | 7ESs.03, 7TWSp.04 AO3 |
| 8(b) | 28 days | 1 | Accept: '1 month' or '1 orbit of the Moon'. | 7ESs.03, 7TWSp.04 AO3 |

# End of Year Test 1

| Qu | Answer | Mark | Further Information | Spec/TWS/AO |
|---|---|---|---|---|
| 1(a) | Solid drawn as tightly packed circles.<br>Liquid drawn as circles that are close together and not touching/only occasionally touching.<br>Gas drawn as circles that are spaced far apart. | 3 | Ignore any other details (e.g. lines indicating movement). | 7Cm.06<br>AO1 |
| 1(b) | Gas to liquid | 1 | | 7Cm.06<br>AO1 |
| 1(c) | fill/occupy<br>compressed/squashed | 1<br>1 | Accept reasonable alternatives. | 7Cm.06<br>AO1 |
| 1(d)(i) | It is not possible to observe particles/particles are too small to be seen. | 1 | | 7TWSp.01<br>AO2 |
| 1(d)(ii) | Expose a balloon to different temperatures/place different balloons at different temperatures.<br>Measure diameter/circumference of the balloons. | 2 | Accept: measure size. | 7TWSp.02, 7TWSp.04<br>AO2 |
| 2(a)(i) | Decomposer | 1 | | 7Be.01<br>AO1 |
| 2(a)(ii) | Recycling of nutrients. | 1 | Accept reasonable alternative wording. | 7Be.01, 7Be.02<br>AO1 |
| 2(b)(i) | The mould covers more of the cake as time passes. | 1 | Accept: the mould is growing. | 7TWSa.02<br>AO2 |
| 2(b)(ii) | Repeat the experiment. | 1 | Accept: use more pieces of cake. | 7TWSc.03, 7TWSa.04<br>AO2 |
| 3(a)(i) | Cell 1 because it has chloroplasts. | 1 | | 7Bs.02, 7Bs.04<br>AO1/2 |
| 3(a)(ii) | Cell 2 because it has root hair/no chloroplasts. | 1 | | 7Bs.02, 7Bs.04<br>AO1/2 |

| Qu | Answer | Mark | Further Information | Spec/TWS/AO |
|---|---|---|---|---|
| 3(b)(i) | Greater/High surface area | 1 | | 7Bs.03 AO1 |
| 3(b)(ii) | Carries oxygen | 1 | | 7Bs.03 AO1 |
| 3(b)(iii) | Contains information/instructions to produce haemoglobin. | 1 | Allow 'nucleus controls the cell's activities'. | 7Bs.03 AO2 |
| 4(a)(i) | Plants only: cell wall + sap vacuole<br>Animals and plants: cell membrane + nucleus + cytoplasm | 2 | | 7Bs.02, 7Bs.04 AO1 |
| 4(a)(ii) | One mark for:<br>Cell membrane – Controls what enters and leaves the cell<br>Cell wall – Keeps the shape of the cell<br>One mark for:<br>Cytoplasm – Site of chemical reactions<br>Sap vacuole – Stores sap | 1<br><br><br><br>1 | | 7Bs.02, 7Bs.04 AO1/2 |
| 4(b) | Agree – Contains cell wall (X) (found in plant cells only).<br>Disagree – Contains nucleus (Y) and mitochondria (Z) (found in both animal and plant cells). | 2 | | 7Bs.02, 7Bs.04 AO2 |
| 5(a) | They cannot be seen with the naked eye/require microscopes to be seen/(usually) single-celled. | 1 | Ignore 'they are small'. | 7Bs.01 AO1 |
| 5(b) | From top to bottom: V, Z, X, W, Y | 2 | 2 marks for all correct.<br>1 mark for three correct. | 7Bp.04, 7TWSc.01 AO2 |
| 6(a) | Iron (nail), tin (foil), copper (wire) | 1 | All three are required for the mark. | 7Cp.05 AO1/2 |
| 6(b) | Plastic rod, cling film, sheet of paper, glass cup | 1 | All four are required for the mark. | 7Cp.05 AO1/2 |
| 6(c) | Good conductor of heat.<br>Good conductor of electricity. | 2 | | 7Cp.05 AO1/2 |

| Qu | Answer | Mark | Further Information | Spec/TWS/AO |
|---|---|---|---|---|
| 7(a) | mixtures<br>different | 1<br>1 | | 7Cp.06<br>AO1 |
| 7(b)(i) | Yes – Lightweight materials make it easier for aircraft to fly.<br>or<br>No – Would not last long. | 1 | Award the mark for a positive or negative answer with the correct reasoning to support it. | 7Cp.06<br>AO1/2 |
| 7(b)(ii) | Yes – Has a longer lifetime.<br>or<br>No – Heavy materials make it harder for aircraft to fly. | 1 | Award the mark for a positive or negative answer with the correct reasoning to support it. | 7Cp.06<br>AO1/2 |
| 7(c) | Carbon | 1 | | 7Cp.06<br>AO1 |
| 8(a) | Kettle<br>Toaster | 2 | | 7Pf.01, 7Pf.02<br>AO1/2 |
| 8(b) | Any object apart from kettle or toaster. | 1 | | 7Pf.01, 7Pf.02<br>AO1/2 |
| 8(c) | Sound | 1 | Not kinetic or movement. | 7Pf.01, 7Pf.02<br>AO1/2 |
| 9(a) | Consists of atoms of silver and sulfur that have been bonded together. | 1 | | 7TWSm.02, 7Cm.04<br>AO1 |
| 9(b) | At low temperatures, it behaves as an insulator (inhibits electron flow).<br>At high temperatures, it behaves as a conductor (allows electron flow). | 2 | | 7Pe.02<br>AO2 |
| 9(c)(i) | Arrowhead pointing anticlockwise on any part of the wire in the circuit. | 1 | | 7Pe.02<br>AO2 |
| 9(c)(ii) | Line with positive gradient, ideally same shape as graph in 9(b). | 1 | | 7Pe.02<br>AO2 |

| Qu | Answer | Mark | Further Information | Spec/TWS/AO |
|---|---|---|---|---|
| **10(a)** | Nitrogen | 1 | | 7ESp03<br>AO1 |
| **10(b)(i)** | Carbon dioxide increases.<br>Oxygen decreases. | 1<br>1 | Accept: water vapour increases. | 7ESp03<br>AO2 |
| **10(b)(ii)** | Nitrogen oxide(s)<br>Sulfur dioxide | 1<br>1 | | 7ESp03<br>AO1 |

# End of Year Test 2

| Qu | Answer | Mark | Further Information | Spec/TWS/AO |
|---|---|---|---|---|
| 1(a) | gravity | 1 | | 7ESs.02 AO1 |
| 1(b) | 330 000/320 = | 1 | | 7ESs.02 AO2 |
| | 1031 | 1 | Accept: 1030, 1032. | |
| 2(a) | Decomposer | 1 | | 7Be.01 AO1 |
| 2(b) | Fungus placed below existing food chain. | 1 | | 7Be.02 AO2 |
| | Four arrows, one pointing from each of the existing organisms in the food chain to the fungus. | 1 | Do not accept three or fewer arrows. | |
| 3(a) | B | 1 | | 7ESp.01 AO1 |
| 3(b) | A | 1 | | 7ESp.01 AO1 |
| 3(c) | Any one from:<br>• tectonic plates float on the mantle<br>• (upper part of) mantle flows. | 1 | Accept: mantle is liquid. | 7ESp.01 AO1 |
| 4(a) | Any one from:<br>• not made of cells<br>• must get into an organism's cell to make copies of itself. | 1 | | 7Bp.02 AO1 |
| 4(b) | Tetanus | 1 | | 7Bp.02 AO2 |
| 5(a) | Element | 1 | | 7Cm.02, 7TWSm.02 AO1 |
| 5(b) | Cr: metal<br>S: non-metal | 1<br>1 | | 7Cm.03 AO2 |
| 5(c) | Scratching | 1 | | 7Cp.05 AO3 |
| 6(a) | Any one of:<br>• it is reflected<br>• it echoes. | 1 | | 7Ps.02 AO1 |
| 6(b) | Arrow starting at seabed and finishing at the hull (bottom) of the ship. | 1 | Accept arrows that do not point vertically upwards, as long as they reach the ship. | 7Ps.02 AO2 |

| Qu | Answer | Mark | Further Information | Spec/TWS/AO |
|---|---|---|---|---|
| 6(c) | 500 ÷ 2 = 250 m | 2 | Allow 1 mark for correct working even if answer is incorrect. Accept 250 (m) for 1 mark without working. | 7Ps.02, 7TWSa.03 AO2 |
| 7(a) | Any one from:<br>• sweep things<br>• move things (along). | 1 | | 7Bs.03 AO1 |
| 7(b) | Respiratory system – Organ system<br>Lung – Organ<br>Group of ciliated cells – Tissue | 2 | 2 marks for all three correct.<br>1 mark for one or two correct. | 7Bs.05 AO1 |
| 8(a) | The centre of the cable must be made of **a conductor**.<br>The outer part of the cable should be made of **an insulator**. | 1<br>1 | | 7Pe.02 AO1 |
| 8(b) | Any one from:<br>• insulator stops electric current from flowing<br>• protects user from electric shocks. | 1 | | 7Pe.02 AO1 |
| 9(a) | Nitrogen | 1 | | 7ESp.03, 7TWSa.05 AO1 |
| 9(b) | Argon | 1 | | 7ESp.03, 7TWSp.03 AO1 |
| 9(c) | 100% − 78% − 21% − 0.05% = 0.95(%) | 1<br>1 | Accept stated percentage, without working for 1 mark. Accept estimate of 1% for 1 mark. | 7ESp.03, 7TWSa.03 AO2 |
| 10(a) | (A space that) contains no particles. | 1 | | 7Pf.04 AO1 |
| 10(b) | Air resistance | 1 | | 7Pf.04 AO1 |
| 10(c) | Any one from:<br>• to make more reliable/improve reliability<br>• to calculate a mean<br>• to identify anomalous results. | 1 | | 7Pf.04, 7TWSc.03 AO3 |
| 11(a) | Neutralisation | 1 | | 7Cc.04 AO1 |

| Qu | Answer | Mark | Further Information | Spec/TWS/AO |
|---|---|---|---|---|
| **11(b)** | Red litmus paper stays red. | 1 | Accept: 'litmus does not change' or similar. | 7Cp.03, 7TWSp.03 AO3 |
| **11(c)** | (Test with) blue litmus paper. | 1 | Accept: 'blue litmus turns pink' or similar. | 7Cp.03, 7TWSc.02 AO3 |
| **11(d)(i)** | Universal Indicator | 1 | Do not accept 'other indicator' or other named indicator. | 7Cp.03, 7TWSc.02, 7TWSa.04 AO3 |
| **11(d)(ii)** | Green | 1 | | 7Cp.03, 7TWSp.03 AO3 |
| **12(a)** | Measures (electric) current. | 1 | Do not accept 'current' on its own. | 7Pe.03 AO1 |
| **12(b)** | Cell: —\|⊢— <br><br> Ammeter: —(A)— | 1 <br><br> 1 | | 7Pe.05, 7TWSm.02 AO2 |
| **12(c)** | Cell | 1 | | 7Pe.04, 7TWSp.04 AO3 |
| **13(a)** | reproduce <br> reproduce | 1 | | 7Bp.03 AO1 |
| **13(b)** | A: Ibis <br> B: Penguin | 1 <br> 1 | | 7Bp.04, 7TWSc.01 AO2 |
| **13(c)** | Any one from: <br> • light/dark front feathers <br> • feet webbed/not webbed <br> • short/long neck | 1 | Accept other clearly distinguishable features. | 7Bp.04 AO2 |
| **14(a)** | Carbon dioxide | 1 | | 7Cp.04, 7TWSa.03 AO3 |
| **14(b)** | Precipitation | 1 | Accept: forms/makes a precipitate. | 7Cc.01 AO1 |
| **14(c)** | Any two from: <br> • wear eye protection <br> • wear protective clothing/lab coat <br> • avoid spillages | 2 | Accept any other appropriate safety precaution. | 7Cc.01, 7Cp.04, 7TWSc.05 AO3 |

| Qu | Answer | Mark | Further Information | Spec/ TWS/AO |
|---|---|---|---|---|
|  | <ul><li>mop up spillages immediately</li><li>keep away from eyes</li><li>keep away from skin.</li></ul> |  |  |  |
| 14(d) | ⚠️ (exclamation mark hazard symbol) | 1 |  | 7TWSm.02, 7TWSc.05 AO3 |

# Periodic Table

**Key**

relative atomic mass
**atomic symbol**
name
atomic (proton) number

| 1 | 2 | | | | | | | | | | | | 3 | 4 | 5 | 6 | 7 | 0 |
|---|---|---|---|---|---|---|---|---|---|---|---|---|---|---|---|---|---|---|
| | | | | | | 1<br>**H**<br>hydrogen<br>1 | | | | | | | | | | | | 4<br>**He**<br>helium<br>2 |
| 7<br>**Li**<br>lithium<br>3 | 9<br>**Be**<br>beryllium<br>4 | | | | | | | | | | | | 11<br>**B**<br>boron<br>5 | 12<br>**C**<br>carbon<br>6 | 14<br>**N**<br>nitrogen<br>7 | 16<br>**O**<br>oxygen<br>8 | 19<br>**F**<br>fluorine<br>9 | 20<br>**Ne**<br>neon<br>10 |
| 23<br>**Na**<br>sodium<br>11 | 24<br>**Mg**<br>magnesium<br>12 | | | | | | | | | | | | 27<br>**Al**<br>aluminium<br>13 | 28<br>**Si**<br>silicon<br>14 | 31<br>**P**<br>phosphorus<br>15 | 32<br>**S**<br>sulfur<br>16 | 35.5<br>**Cl**<br>chlorine<br>17 | 40<br>**Ar**<br>argon<br>18 |
| 39<br>**K**<br>potassium<br>19 | 40<br>**Ca**<br>calcium<br>20 | 45<br>**Sc**<br>scandium<br>21 | 48<br>**Ti**<br>titanium<br>22 | 51<br>**V**<br>vanadium<br>23 | 52<br>**Cr**<br>chromium<br>24 | 55<br>**Mn**<br>manganese<br>25 | 56<br>**Fe**<br>iron<br>26 | 59<br>**Co**<br>cobalt<br>27 | 59<br>**Ni**<br>nickel<br>28 | 63.5<br>**Cu**<br>copper<br>29 | 65<br>**Zn**<br>zinc<br>30 | | 70<br>**Ga**<br>gallium<br>31 | 73<br>**Ge**<br>germanium<br>32 | 75<br>**As**<br>arsenic<br>33 | 79<br>**Se**<br>selenium<br>34 | 80<br>**Br**<br>bromine<br>35 | 84<br>**Kr**<br>krypton<br>36 |
| 85<br>**Rb**<br>rubidium<br>37 | 88<br>**Sr**<br>strontium<br>38 | 89<br>**Y**<br>yttrium<br>39 | 91<br>**Zr**<br>zirconium<br>40 | 93<br>**Nb**<br>niobium<br>41 | 96<br>**Mo**<br>molybdenum<br>42 | [98]<br>**Tc**<br>technetium<br>43 | 101<br>**Ru**<br>ruthenium<br>44 | 103<br>**Rh**<br>rhodium<br>45 | 106<br>**Pd**<br>palladium<br>46 | 108<br>**Ag**<br>silver<br>47 | 112<br>**Cd**<br>cadmium<br>48 | | 115<br>**In**<br>indium<br>49 | 119<br>**Sn**<br>tin<br>50 | 122<br>**Sb**<br>antimony<br>51 | 128<br>**Te**<br>tellurium<br>52 | 127<br>**I**<br>iodine<br>53 | 131<br>**Xe**<br>xenon<br>54 |
| 133<br>**Cs**<br>caesium<br>55 | 137<br>**Ba**<br>barium<br>56 | 139<br>**La***<br>lanthanum<br>57 | 178<br>**Hf**<br>hafnium<br>72 | 181<br>**Ta**<br>tantalum<br>73 | 184<br>**W**<br>tungsten<br>74 | 186<br>**Re**<br>rhenium<br>75 | 190<br>**Os**<br>osmium<br>76 | 192<br>**Ir**<br>iridium<br>77 | 195<br>**Pt**<br>platinum<br>78 | 197<br>**Au**<br>gold<br>79 | 201<br>**Hg**<br>mercury<br>80 | | 204<br>**Tl**<br>thallium<br>81 | 207<br>**Pb**<br>lead<br>82 | 209<br>**Bi**<br>bismuth<br>83 | **Po**<br>polonium<br>84 | **At**<br>astatine<br>85 | **Rn**<br>radon<br>86 |
| **Fr**<br>francium<br>87 | **Ra**<br>radium<br>88 | **Ac**\*\*<br>actinium<br>89 | **Rf**<br>rutherfordium<br>104 | **Db**<br>dubnium<br>105 | **Sg**<br>seaborgium<br>106 | **Bh**<br>bohrium<br>107 | **Hs**<br>hassium<br>108 | **Mt**<br>meitnerium<br>109 | **Ds**<br>darmstadtium<br>110 | **Rg**<br>roentgenium<br>111 | | | | | | | | |

Elements with atomic numbers 112–116 have been reported but not fully authenticated

**La** lanthanoids

**Ac** actinoids

# Glossary

**acid**: substance which has a pH of less than 7 on the pH scale.
**acidic**: having the properties of an acid.
**acid rain**: rainfall that damages plants; it is produced by sulfur dioxide and nitrogen oxides reacting with water in the air.
**adaptation**: feature of something that allows it to do a job (function) or allows it to survive.
**air resistance**: the force that acts to slow down an object moving through air. It varies with the size and shape of the object.
**alkali**: a substance that dissolves in water to make a solution with a pH of more than 7.
**alkaline**: having the properties of an alkali.
**alloy**: a mixture that may contain two or more different metal elements, or one or more metal element and a non-metal element.
**ammeter**: device for measuring current.
**ampere**: unit of electric current, often written as amps, or A.
**amphibian**: vertebrate with moist skin. It lays jelly-coated eggs in water.
**animal kingdom**: kingdom that contains organisms made of more than one cell and are able to move their bodies from place to place.
**anomalous results**: results that don't fit the pattern of the other results obtained.
**arachnid**: arthropod with eight legs and a body in two sections.
**arthropod**: invertebrate with jointed legs and a body in sections.
**atom**: the smallest particle of a substance (element) that can exist and still be the same substance.
**bacterium** (plural bacteria): type of single celled organism that is not a plant or animal or fungus.
**bar chart**: a chart that shows data using columns. They are used to compare different sets of things.
**bird**: vertebrate with feathers. It lays eggs with hard shells.
**blood**: liquid tissue that carries substances around the body.
**blood vessels**: tube-shaped organs that carry blood around the body.
**boiling**: the change of state from liquid to gas.
**boiling point**: the temperature at which a substance changes from a liquid into a gas.
**cell (biology)**: the smallest living part of an organism.
**cell (physics)**: source of energy to make charge move in a circuit.
**cell membrane**: outer layer of a cell that controls what enters and leaves the cell.
**cell wall**: strong outer covering found in some cells (such as plant cells).
**characteristic**: feature of an organism.
**chemical property**: a property that is seen when a substance takes part in a chemical change.
**chemical symbol**: short way of representing an element's name.
**chloroplast**: green part of a cell that makes food using light.
**circulatory system**: group of organs that move blood around the body.
**classification**: arranging things into groups according to similarities and differences.
**compound**: a substance that contains atoms of more than one element strongly held together. Compounds have different properties to the elements they contain.
**conclusion**: decision made using evidence from an investigation.
**condensation**: process in which water vapour particles are cooled and gather to form liquid water droplets.
**conductor**: a material that allows current to flow.
**continent**: large area of land on Earth's surface, separated from other continents by oceans.
**continental drift**: hypothesis that the continents are moving slowly.
**control variable**: variable you keep the same during an investigation.
**core**: the central part of the Earth.
**corrosive**: property of a substance that causes burns to skin and eyes, and damages other materials.
**crust**: hard outer layer of the Earth, made from rock.
**current**: a flow of electric charge.

**cytoplasm**: watery jelly inside cells where new substances are made.
**data**: numbers and words that can be organised to give information.
**decay**: when materials break into smaller parts. Microorganisms often cause this.
**decomposer**: microorganism that causes decay.
**dependent variable**: variable you decide to investigate in an experiment.
**diaphragm**: organ that helps with breathing.
**dichotomous key**: series of choices between two alternative features – used for identifying things.
**digestive system**: group of organs that digest food and move nutrients into the blood.
**dissipate**: to transfer (energy) in a wasteful way.
**earthquake**: event where the sudden movement of tectonic plates causes the crust to move and vibrate violently.
**echo**: a reflected sound.
**electron**: subatomic particle which carries negative charge.
**element**: substance that contains only one type of atom.
**evaluate**: look at the good and bad points of something in order to make a decision.
**evaporation**: process in which water particles gain enough energy to escape from the surface of liquid water and form vapour.
**evidence**: data or observations we use to support or contradict an idea.
**excrete**: get rid of wastes made inside an organism.
**fault line**: crack in Earth's crust caused by movement of tectonic plates.
**fertile**: able to reproduce and have offspring.
**fish**: vertebrate with slimy scales. It lays eggs in water.
**flowering plant**: type of plant that produces flowers.
**food chain**: diagram showing feeding relationships in a habitat – each species is a food source for the species at the next level up.
**food web**: diagram to show how food chains interconnect in a habitat.

**formula**: shows the chemical symbols of elements in a compound, and how many of each type of atom there are.
**fossil fuel**: fuel such as coal, oil or gas that is produced naturally over millions of years from the remains of dead animals and plants.
**free fall**: the speeding up downwards motion of a body due to its weight.
**freezing**: the change of state from liquid to solid.
**friction**: force between two surfaces that are pressed together. It acts to stop the surfaces sliding over one another.
**function**: another word for 'job'.
**fungus** (plural fungi): type of microorganism that reproduces by producing spores.
**gas syringe**: device used to measure a volume of gas.
**gravitational field strength** ($g$): a measure of how much force gravity exerts on each kilogram of mass.
**gravity**: force between two objects that is due to their mass.
**greenhouse gas**: gas that traps heat and keeps the Earth warm.
**habitat**: the place where an organism lives.
**haemoglobin**: substance that carries oxygen.
**hazard**: harm that something may cause.
**hazard symbol**: symbol which warns you about the dangers of an object, substance or radiation.
**heart**: organ that pumps blood through blood vessels.
**heating**: transfer of energy that causes a temperature rise or is due to a temperature difference.
**high tide**: when the sea reaches its highest level relative to the land.
**host cell**: the cell that a virus needs to get into in order for it to replicate.
**hybrid**: offspring produced by reproduction between two different species.
**hypothesis** (plural hypotheses): a statement or claim that can be tested using experiments. It is proved or disproved by scientific enquiry.
**independent variable**: variable you decide to change in an experiment.

**indicator**: chemical that changes colour in an acid or alkali.
**inertia**: tendency of a moving or stationary object to resist changes in its motion.
**insect**: arthropod with six legs and a body in three sections.
**insoluble**: a substance that does not dissolve.
**insulator**: a material which does not allow current to flow.
**invertebrate**: animal without a skeleton inside it and without a 'backbone'.
**kidneys**: organs that remove wastes from the blood to produce urine.
**kinetic**: due to motion.
**kingdom**: the biggest of the groups that scientists use to classify organisms.
**large intestine**: organ that absorbs water from undigested food.
**law of gravitation**: mathematical theory developed by Newton that allows scientists to calculate the motion of planets around the Sun, and moons around planets.
**life process**: something that all living things do.
**limewater (calcium hydroxide)**: clear and colourless liquid that turns milky when carbon dioxide is added.
**limitation**: feature of a scientific idea or theory that either is not clearly supported by evidence, or that cannot yet be explained.
**line graph**: graph that shows data points plotted on a grid and connected with lines. Line graphs are often used to show how one thing changes with time. Time is put on the horizontal axis.
**litmus**: type of indicator which turns red in an acid and blue in an alkali.
**liver**: organ that makes and destroys substances.
**low tide**: when the sea reaches its lowest level relative to the land.
**lunar eclipse**: when the Earth blocks the light of the Sun from reaching the Moon.
**lungs**: organs that get oxygen into the blood and remove carbon dioxide.
**magnification**: the amount to which something is magnified.
**malleable**: can be formed into different shapes.
**mammal**: vertebrate that usually gives birth to live young (except for the platypus and echidna) and that feeds its young on milk.
**mantle**: hot, flowing rock between the core and the crust.
**mass**: the amount of matter in an object – it is measured in grams or kilograms.
**medium**: the substance a sound wave travels through.
**melting**: the change of state from solid to liquid.
**melting point**: the temperature a substance melts at, and changes from a solid into a liquid.
**microorganism**: tiny organism. We must use a microscope to see it.
**microscope**: piece of equipment that magnifies very small things.
**mitochondrion** (plural mitochondria): part of a cell where respiration happens, which releases energy.
**mixture**: two or more elements or compounds mixed together. They can easily be separated.
**model**: simple way of showing or explaining a complicated object or idea.
**moon**: natural object that orbits a planet.
**mould**: fungus that decays things.
**multicellular**: made of many cells.
**neap tide**: least high tide, because the force of gravity due to the Sun is at right angles to the force of gravity due to the Moon.
**nervous system**: group of organs that control the senses of the body.
**neutral**: neither acid nor alkali. If soluble, it produces a solution of pH 7.
**neutralisation**: chemical reaction between an acid and an alkali which produces a neutral solution.
**nucleus**: control centre of a cell.
**nutrient**: a substance that an organism needs to stay healthy and survive.
**nutrition**: the process of getting substances needed for survival.
**offspring**: new organism made when parents reproduce.
**orbit**: circular or nearly circular path that a planet follows around the Sun, or a moon follows around its planet.

**organ**: part of an organism that has an important job (function). Organs are made up of different tissues.
**organism**: living thing.
**organ system**: group of organs working together.
**palisade cell**: cell found in plant leaves, which contains many chloroplasts.
**parallel circuit**: an electrical circuit made up of more than one loop.
**partial lunar eclipse**: when the Earth's shadow partly covers the Moon.
**partial solar eclipse**: when the Moon partly blocks light from the Sun from falling on the Earth.
**particle model**: model that describes how particles are arranged differently in solids, liquids and gases.
**penumbra**: area of partial shadow.
**Periodic Table**: list of all the elements, in an order.
**pH scale**: scale from 0 to 14 which measures how strong or weak an acid or alkali is.
**physical property**: the property that can be observed or measured without changing the basic nature of the substance.
**planet**: object that orbits a star and which is large enough to have cleared its orbit of all smaller objects.
**plant kingdom**: kingdom that contains organisms made of more than one cell and make their own food.
**plate boundary**: the area where two tectonic plates meet, forming a line.
**pollutant**: substance produced by human activities that may damage the environment or animals living in the environment.
**precipitate**: insoluble solid formed when soluble substances react together.
**precipitation**: process in which water droplets in the air grow large enough that they fall to the ground as rain, sleet, snow or hail.
**prediction**: what you think will happen in an investigation.
**primary consumer**: animal that eats plants (producers). These animals are herbivores. For example, an antelope eats grass.
**producer**: organism that makes its own food, such as a plant.

**product**: substance made during a chemical reaction.
**pure**: substance that contains only one element or compound.
**range**: the difference between the highest and lowest values in a set of data.
**reactant**: substance that changes in a chemical reaction to form products.
**reflection**: when something hits a surface and 'bounces back' towards its source.
**replication**: the process in which a host cell makes copies of a virus inside it.
**reproduce**: when organisms have young (or offspring).
**respiration**: chemical process that happens inside cells to release energy.
**respiratory system**: group of organs that get oxygen into the blood and remove carbon dioxide. Also called the breathing system.
**risk**: chance of a hazard causing harm.
**root hair cell**: plant cell found in roots that is adapted for taking in water quickly.
**scientific method**: stages that scientists use to test out their ideas.
**scientific question**: question that scientists can answer using an experiment.
**secondary consumer**: animal that eats a primary consumer. These animals are carnivores. For example, a hyena eats an antelope.
**sensitivity**: how an organism detects changes in things inside and around it.
**series circuit**: an electrical circuit made of just one loop.
**skeletal system**: group of organs that support the body and allow movement.
**skin**: organ that protects the body and helps it sense things.
**small intestine**: organ that digests food and absorbs it into the blood.
**solar eclipse**: when the Moon blocks light from the Sun from falling on the Earth.
**Solar System**: the Sun and all the other objects that move around it.
**soluble**: substance that dissolves to form a solution.
**solution**: a mixture of a soluble substance and a liquid.
**sound wave**: the way that sound passes through a substance by alternately pushing and pulling particles to make them vibrate.

**specialised cell**: cell with adaptations for a certain job.
**species**: a group of very similar organisms that can reproduce to produce fertile offspring.
**spore**: single cell released into the air by a fungus that is able to grow into a new fungus.
**spring tide**: highest high tide, because the force of gravity due to the Sun is in the same direction as the force of gravity due to the Moon.
**state of matter**: the three forms that a substance can exist in: solid, liquid and gas.
**stomach**: organ that helps to digest food.
**surface area**: the area of a surface, measured in squared units such as square centimetres ($cm^2$).
**tectonic plate**: area of the Earth's crust that is moving.
**tertiary consumer**: animal that eats a secondary consumer. These animals are carnivores. For example, a crocodile eats a hyena.
**theory**: scientific idea that explains a number of different observations and which has been tested by evidence.
**thermal**: linked to temperature.
**tissue**: group of cells of the same type.
**total lunar eclipse**: when the Earth's shadow completely covers the Moon.
**total solar eclipse**: when the Moon completely blocks light from the Sun from falling on the Earth.
**umbra**: area of total shadow.
**unicellular**: one cell. Used to describe a single-celled organism.

**Universal Indicator**: type of indicator that can change into a range of colours depending on whether the solution is acidic or alkaline and how strong it is.
**useful energy**: energy that is transferred to make something useful happen.
**vacuole**: storage space inside some cells (such as plant cells).
**vacuum**: a space that contains no particles.
**variable**: something you can measure or observe.
**variation**: differences between characteristics.
**vertebrate**: animal with a skeleton inside it, including a 'backbone'.
**vibration**: when something moves back and forwards many times, we say it vibrates.
**virus**: particle that can only replicate (copy itself) when it is inside a living cell.
**volume**: how much space a substance takes up. Measured in $cm^3$ or litres.
**wasted energy**: energy that transfers in a way or to a place where we cannot easily use it.
**water cycle**: the sequence where water evaporates from the oceans, condenses to form clouds, falls as precipitation and returns to the oceans.
**weight**: the size of the pull of gravity for a given mass. You can feel something's weight when you try to stop it from falling.
**wilt**: when a plant droops because it does not have enough water.
**word equation**: model showing what happens in a chemical reaction, with reactants on the left of an arrow and products on the right.
**work**: transfer of energy by the action of a force to make an object move.